7th Symposium on Vacuum-based Science and Technology

(SVBST 2013)

Journal of Physics: Conference Series Volume 564

Kolobrzeg, Poland
19-21 November 2013

ISBN: 978-1-63439-672-1
ISSN: 1742-6588

Printed from e-media with permission by:

Curran Associates, Inc.
57 Morehouse Lane
Red Hook, NY 12571

Some format issues inherent in the e-media version may also appear in this print version.

Printed by Curran Associates, Inc. (2015)

For permission requests, please contact the Institute of Physics
at the address below.

Institute of Physics
Dirac House, Temple Back
Bristol BS1 6BE UK

Phone: 44 1 17 929 7481
Fax: 44 1 17 920 0979

techtracking@iop.org

Additional copies of this publication are available from:

Curran Associates, Inc.
57 Morehouse Lane
Red Hook, NY 12571 USA
Phone: 845-758-0400
Fax: 845-758-2634
Email: curran@proceedings.com
Web: www.proceedings.com

Table of contents

Volume 564

7th Symposium on Vacuum-based Science and Technology (SVBST2013)
19–21 November 2013, Kołobrzeg, Poland

Accepted papers received: 11 November 2014
Published online: 28 November 2014

These are the proceedings of the 7th **Symposium on Vacuum based Science and Technology** organized in Kołobrzeg (PL) on November 19-21, 2013 by the Institute of Technology and Education, Koszalin University of Technology and the Clausius Tower Society under auspices of the Polish Vacuum Society (PTP) and the German Vacuum Society (DVG) and in collaboration with the BalticNet PlasmaTec and the Society of Vacuum Coaters (SVC). It was accompanied by the 12-th Annual Meeting of the German Vacuum Society.

The mission of the Symposium is to provide a forum for presentation and exchange of expertise and research results in the field of vacuum and plasma science. After already six successful meetings organized alternately in Poland and Germany our goal is to continue and foster cooperation within the vacuum and plasma science community.

This year, the **Rudolf-Jaeckel Prize**, awarded by the DVG for outstanding achievements in the field of vacuum based sciences, was presented to Dr Ute Bergner, president of the VACOM Vakuum Komponenten & Messtechnik GmbH and a member of our community.

The full-day course organized in the framework of the Educational Program by the Society of Vacuum Coaters (SVC) and entitled: An Introduction to Physical Vapor Deposition (PVD) Processes was held on November 18, 2013 as a satellite event of the Symposium. The instructor was Prof. Ismat Shah from Delaware University (US).

The Clausius Session, already traditionally organized during the Symposium was addressed this year to young generation. We invited our young colleagues to attend a series of educational lectures reporting on achievements in graphene science, scanning probe microscopy and plasma science. Lectures were given by: Prof. Jacek Baranowski from the Institute of Electronic Materials Technology in Warsaw, Prof. Teodor Gotszalk from the Wroclaw University of Technology and Prof. Holger Kersten from the Christian Albrechts University in Kiel.

The Symposium was accompanied by an industry exhibition attended by the representatives of leading companies offering vacuum equipment, complete solutions for plasma based technology as well as advanced research equipment.

Witold Gulbiński
Michael Kopnarski
Frank Richter
Jan Walkowicz

Journal of Physics: Conference Series **564** (2014) 011002 doi:10.1088/1742-6596/564/1/011002

Peer review statement

All papers published in this volume of *Journal of Physics: Conference Series* have been peer reviewed through processes administered by the proceedings Editors. Reviews were conducted by expert referees to the professional and scientific standards expected of a proceedings journal published by IOP Publishing.

Hydrogenated amorphous carbon coatings on implants drastically reduce biofilm formation and water permeation

Falk Bernsmann[1], Norbert Laube[1*], Gerhard Baldsiefen[2] and Mattia Castellucci[3]

[1] NTTF Coatings GmbH, Maarweg 32, D-53619 Rheinbreitbach, Germany
[2] Institute of Thin Film Technology, Kaiserslautern University of Technology, Maarweg 32, D-53619 Rheinbreitbach, Germany
[3] Università degli Studi di Trieste, piazzale Europa 1, I-34127 Trieste, Italy

E-mail: norbert.laube@nttf-coatings.de

Abstract. Inflammations and crystalline bacterial biofilms (encrustations) remain a major complication in long-term artificial urinary tract drainage. To solve this problem we present urological implants with coatings made of amorphous hydrogenated carbon (a-C:H) that show excellent protection from encrustation in-vitro as well as in-vivo. Part of the success of a-C:H coatings is attributed to their ability to act as a diffusion barrier between an implant and the body, which prevents leaching of solvents from polymeric implants. To further enhance their barrier properties a-C:H coatings are combined with parylene coatings to develop diffusion-barrier multilayer coatings with a total thickness between 0.2 μm and 0.8 μm. The combination of the two types of coatings leads to a reduction of water diffusion by a factor of up to ten with respect to uncoated 25 μm thick polyimide sub-strates. The diffusion of water vapour from a controlled atmospheric pressure chamber through coated foils to a vacuum chamber is measured in a custom-built device.

1. Introduction

1.1. Modifications on implants and other biomedical devices

Thanks to the progress of medical technology, more and more sophisticated implants like in-situ blood-pressure sensors, cardiac pacemakers or retinal implants are developed. Also "routine" implants such as indwelling urological stents and catheters, the latter in principle in use since ancient times [1], have been undergoing continual improvement since that time. The limiting factors on implantation times are governed by the (undesired) interactions of the implant surface with the body and vice versa. In particular the build-up of bacterial and/or crystalline biofilms or mucus deposits, as well as the attachment of debris, often leads to the need of early implant removal. Furthermore, many implant bulk materials, for example silicon or polyurethane, do not withstand the corrosive body environment, or they can leach toxic substances, such as heavy-metal ions or plasticisers. In both examples, improvements of the implant's biocompatibility can only be achieved with an optimization of the implant/body interface: in the specific cases by prevention of bacteria adhesion and by encapsulation (**Table 1**). Tailor-made thin film coatings, which neither alter the implant's surface topography to an undesired extent nor hinder its functionality, are needed. Concerning encapsulation traditional polymers and even common high barrier coatings like poly-p-xylylenes (parylenes) are often not sufficient to provide the necessary diffusion-tightness.

* To whom any correspondence should be addressed.

Table 1: Properties of implants and other medical devices that can be adapted by coatings to complement the bulk materials' properties.

	Task	Problem Area (Examples)		Example of Use
Chemical Properties	Diffusion barrier	Irritations and inflammations due to migration of substances out of plastic materials / polymers		Urological implants
		Entry of (body-)fluids into electrically active implants		Neuroimplants
		Avoidance of uncontrolled migration of reactive materials		Prostate seeds
	Corrosion prevention	Sterilization of metallic medical devices		Increase of life-time of, e.g., surgical instruments and endoscopes
	Control of adhesion of bacteria, cells, proteins and biominerals	Biofilm formation and growth		Cardiological and urological implants Contact lenses with low protein adhesion
	Functionalization with biomolecules	Active molecules no longer have to be incorporated into the basic (bulk) material (e.g. polymers)		Coatings of implants with growth- and adhesion factors to stimulate tissue formation (orthopedics, trauma surgery, dentistry) Generation of anchor molecules for the binding of locally acting drugs (e.g. „drug-eluting" stents, chitosan) Fabrication of anti-bacterial but non-cytotoxic surfaces on urological catheters and stents
Physical Properties	Increase of resistance of wear and tear, hardness	Early implant- or instrument failure Wear debris gets into bloodstream / body		Orthopedic and dental implants Increase of scratch resistance of optical lens systems
	Optimizing of tribological properties (coefficient of friction)	Friction too high or too low		Gliding properties of implants and implant-guide wires
	Variation of topography	(Micro-)Roughness Adhesion of unwanted and non-adhesion of desired substances or microorganisms, respectively		Nano-structuring of dental implants for improvement of osseointegration
	Adjustment of wettability	Adhesion of unwanted and non-adhesion of desired substances, resp.	Hydrophilicity	„Super-hydrophilic" surfaces allow for instantaneous wetting of the implants with blood, thereby significantly improving e.g. osseointegration
			Hydrophobicity	Lotus effect, „easy-to-clean"-surfaces
	Optical properties	Reduced (micro-)roughness		Antireflection- and antifogging-coatings on optical lenses
	Electrical conductivity	Non-biocompatible electrodes		Cardiac pacemakers, neuroimplants
	Radioactivity	Local radiotherapy		Radiotherapeutic procedures at locally limited cancer
Biological Properties	Increase of body-compatibility	Implantat rejection / failure		Vessel supports (stents)
	Hemocompatibility	Adhesion of blood components, blood coagulation		Artificial heart valves, dialysis-tubes, bone implants
	Cell adhesion	Adhesion		Burn wound covering
	Antibacterial activity	(Implant-associated) Infections		Implant coatings with photo-catalytically active titanium dioxide

1.2. Crystalline biofilm formation: a common problem in urology

Urologic indwelling catheters and especially ureteral stents are among the most frequently used medical implants. A wide spectrum of indications for artificial urinary drainage exists in order to ensure unobstructed transport of urine, which can be compromised for example due to a tumour applying a confining pressure on the ureter (**Figure 1**).

However, unpleasant side-effects of artificial urinary drainage are frequent. In most cases they are triggered by bacteria entrained into the urinary tract already during catheter placement or migrating on the inner or outer part of the indwelling implant's surface. In fact, most patients with an indwelling urinary catheter for 30 days or longer develop bacteriuria [2]. Complications of intermediate (2-6 weeks) and long-term (> 6 weeks) catheterization include pain, urgency, discomfort and bladder irritability (bladder spasms and contraction) as well as chronic renal inflammation, chronic pyelonephritis, nephrolithiasis, cystolithiasis, recurrent (mostly) polymicrobial urinary tract infections with symptomatic pyelonephritis, bacteraemia, sepsis and death [3, 4, 5, 6, 7, 8, 9].

These serious complications can be caused by the formation of organic biofilms and deposition of inorganic salts on the implants' surfaces thereby often provoking fatal functional failure. Within only a few hours in-situ, a conditioning film composed of urinary proteins forms, and subsequently bacteria start to attach to the surface of the implants. After only a couple of days the surfaces of the urinary stents and catheters are colonized by bacteria forming an infectious bacterial biofilm. Consecutive deposition of insoluble inorganic salts - as a consequence of bacteria-metabolism-related rise in urinary pH - leads to the formation of sharp-edged crystals (**Figure 2**).

A bacterial biofilm can be considered as a chronic urinary tract infection, which, under varying antibiotic therapy, may easily result in therapeutically hard to reach multi-resistant bacterial strains. For this reason the catheters and stents have to be replaced at the first indication of complications. In the extreme case, the intracorporeal indwelling time of a standard implant may be only a few days. Early replacement is unpleasant to the patients and produces high economic costs. In the US, the overall annual cost for medical intervention of catheter-associated urinary tract infections alone amount to $424 million to $451 million [10] and catheter-associated bacteraemia is estimated to cost approximately $2,900 per episode [11].

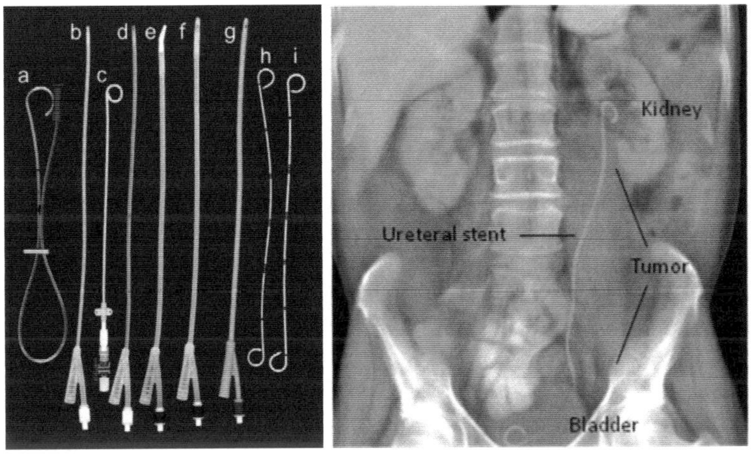

Figure 1: Left: Typical urological implants for different use (*a*: suprapubic nephrostomy catheter; *b*, *d*: suprapubic bladder catheters; *c*: nephrostomy catheter; *e*, *f*, *g*: transurethral bladder catheter; *h*, *i*: ureteral stents). Right: Ureteral stents are implanted in the ureter between kidney and bladder to ensure unrestricted urinary outflow (computer tomography (CT) showing a large ureter-confining tumour requiring permanent stent placement).

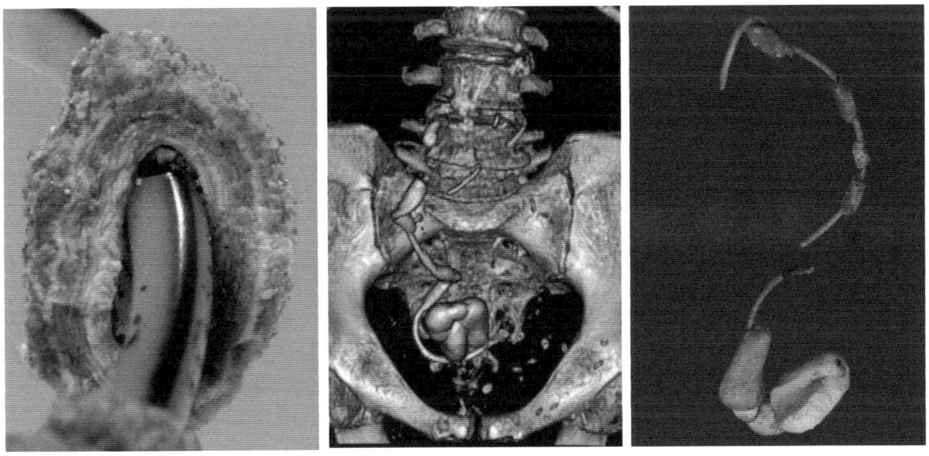

Figure 2: From left to right: Typical confining encrustation of an ureteral stent after only a few weeks of indwelling, "forgotten" ureteral stent with huge mineral deposition (approx. 150 g in total) as an extreme example of crystalline bacterial biofilm (CT-view of the in-situ situation and explanted stent). The embrittled stent broke during explantation. Crystalline material on both stents is composed of Mg- and Ca-phosphates (90 % carbonate (hydroxyl-)apatite and 10 % struvite, and 50 % carbonate (hydroxyl-)apatite and 50 % struvite, respectively).

1.3. Hydrogenated amorphous carbon coatings

Amorphous carbon (a-C), also called diamond-like carbon (DLC) is a class of materials with a wide range of properties from graphitic to diamond-like that can be tuned by the carbon hybridisation and hydrogen content. DLC are generally characterised by good tribological properties, high chemical resistance and biocompatibility [12]. The typically 10-200 nm thick DLC-surface coatings described in this work were especially developed for medical and biological applications and belong to the subgroup of hydrogenated amorphous carbon, a-C:H (approximately 20 at% hydrogen). Due to the processing temperatures below 50 °C, the a-C:H coatings are particularly suitable for flexible polymer-based medical implants; the maximum operating temperature is 600 °C. Despite a Vickers hardness of 1,000-3,000 HV and a friction coefficient against steel as low as 0.05, the surface conform coating is still very flexible and does not detach even after extreme flexing.

2. Experimental Part

2.1. Surface preparation

Different hydrogenated amorphous carbon coatings were deposited on polyurethane (PU) tubes taken from ongoing production of a stent manufacturer by plasma-enhanced chemical vapour deposition (PECVD) using a radiofrequency-(RF)-driven (13.56 MHz), inductively coupled plasma beam source (CCR COPRA®; stainless steel, 266 mm diameter) with a transverse magnetic field providing a high density plasma [13, 14, 15]. Deposition conditions are characterized by low temperature and low gas pressure (temperature approximately 25 °C and pressure of 0.1 Pa to 0.5 Pa). The film forming particle flux is adjusted by varying total gas pressure and high frequency power, and it is monitored by faraday cup measurement. With ion energies between 15 eV and 25 eV and ion current densities between 0.1 $\mu A/cm^2$ and 0.2 $\mu A/cm^2$, the energy flux density is reduced by two orders of magnitude with respect to the conditions typical for the production of diamond-like carbon films [16]. This allowed not only for a compact structure (Vickers hardness $H \approx 10$ GPa) of the amorphous layer, but also excellent elastic properties, preventing the stripping off of a layer from the highly flexible stent material (Youngs modulus $E \approx 60$ GPa) [17]. The ratio E/H of 6 is in agreement with the constraint model of Angus et al. [18]. That means the carbon films are thermodynamically stable. The film forming parameter settings ensured a moderate layer growth at the lowest possible thermal load of the substrate.

In total, six different a-C:H coatings were tested against an uncoated polyurethane reference (B-PU). The thin films a-C:H-2, a-C:H-125 and a-C:H-Uro were deposited each from acetylene (C_2H_2-flow ≈ 30 sccm) precursor gas at different gas pressures (≈ 0.07 Pa, ≈ 0.15 Pa, ≈ 0.15 Pa) and RF-powers (150 W, 125 W, 250 W). In order to achieve different surface wettabilities the gases O_2, NH_3 and CF_4 (gas flow ≈ 30 sccm, each; RF-power = 125 W), respectively, were added during the coating process for the deposition of the other three thin films a-CO_x-H, a-C:H-NH_3 and a-C:H-F-2.

Multilayer coatings were deposited on polyimide foils (DuPont Kapton HN, nominal water vapour transmission rate $WVTR = 54$ g/day/m² at 37 °C) of 5 cm diameter and 25 μm thickness. Hydrogenated amorphous carbon coatings of 20 nm thickness were deposited using the same system mentioned before but with a different procedure: First, the samples were exposed to an oxygen plasma for 10 min at a pressure of 2 Pa and a RF power of 200 W for surface cleaning and activation. The a-C:H layers were deposited from acetylene (C_2H_2) within 10 min at a pressure of 0.7 Pa and a power of 107 W. Parylene C was deposited from 2-chloro-[2,2]-paracyclophane using a Gorham process [19] with a chamber pressure of 3 Pa, a vaporiser temperature of 150 °C and a reactor temperature of 720 °C.

The coating thickness was measured on silicon slides coated in the same processes as the polyimide foils either with an ellipsometer (Dr. Riss Ellipsometerbau EL X-02C, 633 nm wavelength, variable angle of incidence) for a-C:H coatings or with a profilometer (Veeco Dektak[3]) for parylene coatings.

2.2. In-vitro experiments for determination of the encrustation tendency

The encrustation tendencies of the seven different surface types were tested in several runs. Each run consisted of eight coated (two coating types, four samples each) and four uncoated reference samples.

The samples were simultaneously incubated under defined reproducible conditions in the extended in-vitro crystallization model. This so-called "encrustator" device is a water bath thermostated flow-crystallizer model according to the mixed suspension mixed product removal method (MSMPR) [20] and it allows the simultaneous investigation of twelve test samples in artificial or native urines. A detailed description of the experimental procedure can be found elsewhere [21]. For better reproducibility of the experiments, the systematic investigations presented here were performed using synthetic urines composed of dissolved salts and few low-molecular substances [22].

The common urinary tract infection is induced by urease-producing microorganisms. As urease splits urea into ammonia and CO_2, a typical pH rise in the urine is observed. The increase in the urine's pH provides the supersaturation required for the formation of phosphate salts. In-vitro, this pH increase was triggered by direct addition of urease to the artificial urine. Two partial solutions, A (containing, inter alia, urea and albumin) and B (containing, inter alia, urease), were each fed at a constant rate into the encrustator's reaction chamber and mixed; at this moment, a potentially crystal forming solution exists. The constant addition of fresh urine avoids depletion of one of the lithogenic substances during incubation and maintains a quasi-constant urinary supersaturation within the reactor, allowing for salt formation over the entire incubation time. An overflow keeps the urine volume in the reaction chamber constant.

After 45 hours incubation time, morphology, amount and composition of the precipitations were determined. The salts obtained on the different stent surfaces were dried at room temperature (24 h), photographically documented, and then dissolved in 10 mL 1 N hydrochloric acid.

From the solution's Ca, Mg, and PO_4^{2-} concentrations, a qualitative and quantitative analysis of the crystalline material was performed, using a procedure previously evaluated by infrared spectroscopy and routinely used for geochemical rock characterization. By applying mass balance calculations, a standardized mineral paragenesis of the dissolved salts was evaluated. This computed mineralogical composition consists of the Ca-phosphates carbonate (hydroxyl-)apatite and brushite, and of the Mg-ammonium-phosphate, struvite. These salts were determined in a number of investigations to represent the most abundant minerals found in biofilms formed on ureteral stents. They occur either as single phases or in paragenesis. Their (ideal) composition is assumed to be $Ca_{10}(PO_4)_5CO_3(OH)_2$, $CaHPO_4 \times 2H_2O$, and $MgNH_4PO_4 \times 6H_2O$, respectively [23, 24, 25].

For each run, a normalized amount of minerals for each surface type was calculated by dividing the molar concentrations of all minerals found on each surface type by those values found for the uncoated PU surface. This allows for both qualitative and quantitative analysis of the crystalline material.

2.3. Surface energy measurement

The surface properties determined in this study were motivated by the general ideas and thermodynamic models on the biofilm and mineralization determining surface properties [26, 27, 28] claiming hydrophilic/hydrophobic interactions as one of the key factors of bacterial adhesion and crystallization [29]. The surface free energies of the samples were determined according to the method of Owens, Wendt, Kaelble and Rabel [30] by optical contact angle measurements with three different test liquids (water, diiodomethane, ethylene glycol) using the sessile drop method and an OCA 15 plus (DataPhysics Instruments GmbH, Filderstadt, Germany) as described in [21].

2.4. Diffusion measurement

We have developed a device to measure the diffusion of various gases from a controlled environmental chamber through thin foils to a high vacuum chamber, where the diffused species were detected by a quadrupole mass spectrometer (Hiden Analytical Halo 201). The principle of the device is sketched in **Figure 3**. The sample foil was mounted on a perforated titanium sheet (0.1 μm thickness, 500 holes of 70 μm diameter) to avoid deformation and damage due to the pressure difference. The temperature of the sample and the environmental chamber were controlled with a computer program (National Instruments LabVIEW) by ceramic heaters. The humidity in the environmental chamber was automatically adjusted by the inflow of wet or dry gas.

Before starting a measurement the environmental chamber was flooded with argon and the high vacuum chamber was evacuated to a water partial pressure (p_{H2O}) of about 3 μPa. Once p_{H2O} has stabilised, the sample temperature was set to the desired value and the environmental chamber was filled with air at ambient pressure and 100 % relative humidity. An exemplary record of p_{H2O} during an experiment is shown in **Figure 4**.

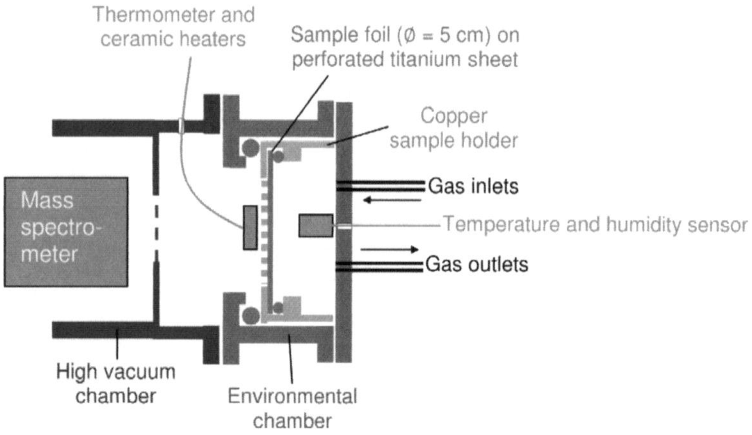

Figure 3: Sketch of the diffusion-measurement device built for this work.

Depending on the temperature, p_{H2O} stabilises at different equilibrium values. The difference in p_{H2O} (Δp_{H2O}) between the equilibrium pressure and the background pressure measured at the same temperature without water in the environmental chamber was supposed to be proportional to the water vapour transmission rate (*WVTR*). The value of Δp_{H2O} measured with an uncoated polyimide foil of known *WVTR* was used to obtain a calibration factor for calculating the *WVTR* of coated foils from their Δp_{H2O}. Once the *WVTR* of uncoated ($WVTR_f$) and coated ($WVTR_{cf}$) foils was known, the *WVTR* of the coating itself ($WVTR_c$) could be calculated as

$$\frac{1}{WVTR_c} = \frac{1}{WVFR_{cf}} - \frac{1}{WVTR_f}$$

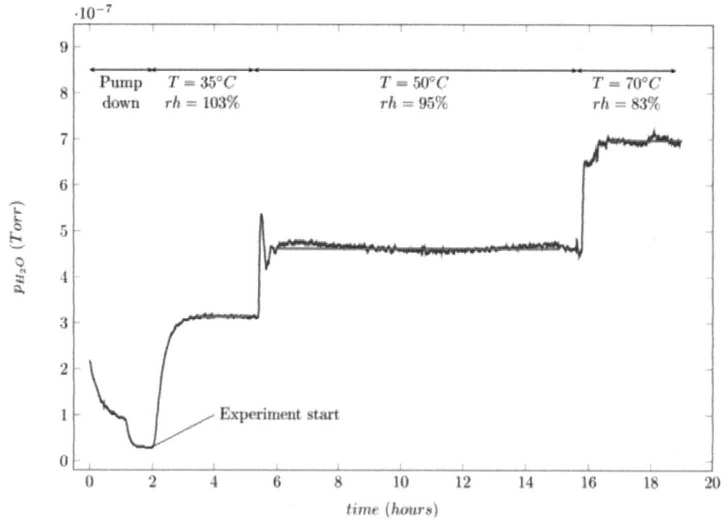

Figure 4: Partial pressure of water diffusing through a polyimide foil coated with 20 nm of a-C:H and 800 nm of parylene at increasing temperatures. The blue line represents the recorded values; the red lines represent the equilibrium pressures used to calculate the water vapour transmission rate.

3. Results

3.1. Surface free energy and crystal formation

Figure 5 and **Figure 6** show the results obtained from the in-vitro crystallization experiments and those of the determinations of the free surface energies, respectively. The positive effect of any of the different a-C:H film compositions on the encrustation tendency is clearly shown. All types of a-C:H films tested show less encrustations than the commercially available uncoated "premium stent" ("B-PU") claimed being optimized for "long-term indwelling", which is taken as reference. According to the approach of [29], the slightly hydrophilic a-C:H-films (water contact angles between 53.3° (± 0.5°) and 69.2° (± 0.4°) compared to 30.9° (± 0.6°) for uncoated B-PU) are in particular suitable to hinder the adherence of microorganisms with "hydrophobic" properties.

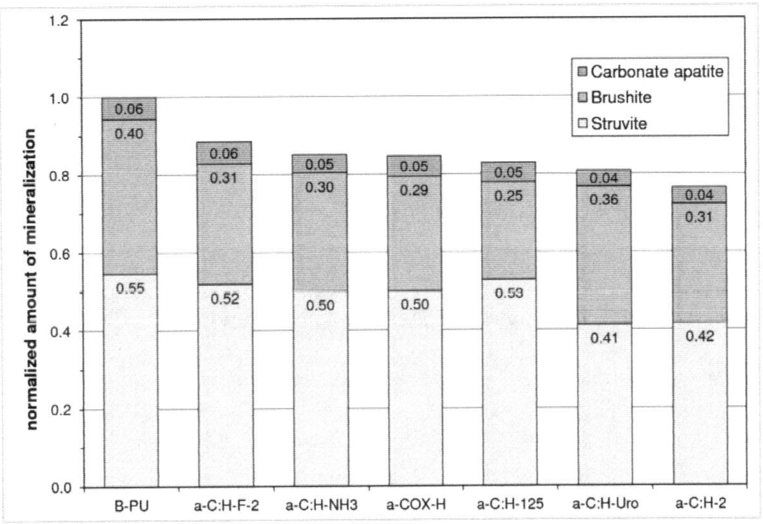

Figure 5: Different surface types vs. normalized amount of phosphate salts found on these surfaces after 45 hours of incubation. The calculated fractions of the standardized mineral content are also given.

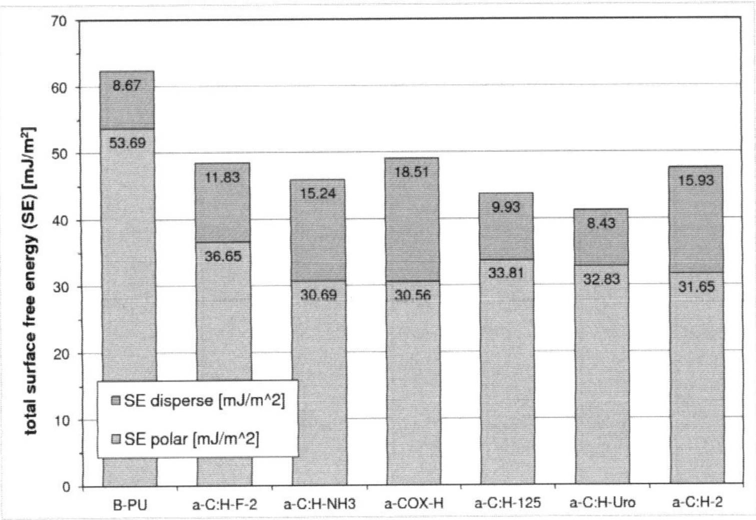

Figure 6: Different surface types in order of decreasing mineral amounts vs. surface free energy. The total surface free energies as well as their polar and disperse fractions are shown for each examined coating.

3.2. Water vapour permeation

Figure 7 shows the *WVTR* of polyimide foils coated with layers of hydrogenated amorphous carbon and parylene of two different thicknesses measured at foil temperatures from 35 °C to 70 °C. The two samples labelled "DLC" are a-C:H coatings deposited by the same process in two separate batches. The a-C:H coatings on top of the two thicknesses of parylene are also deposited in two separate batches. Since there are only slight differences in the properties of the a-C:H coatings from different batches, all of them are labelled "DLC" in **figure 6** and **figure 7** and supposed to be equivalent for the results' interpretation.

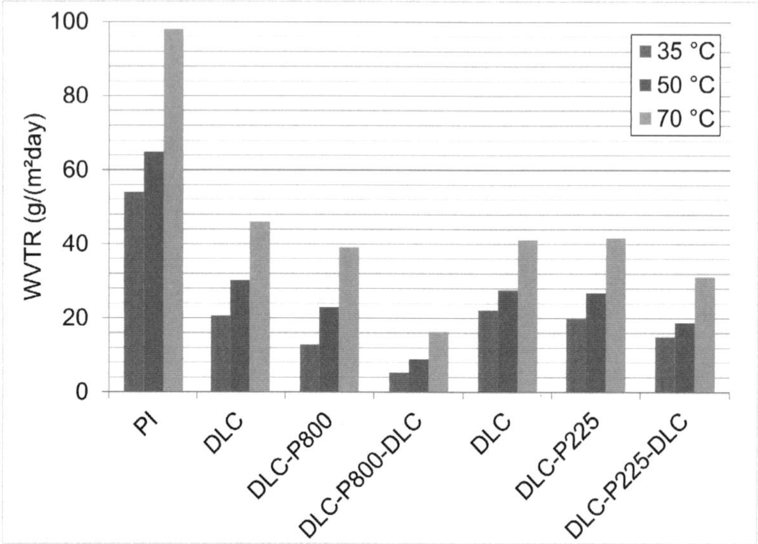

Figure 7: *WVTR* of polyimide (PI) coated with layers of hydrogenated amorphous carbon (DLC) and parylene with 800 nm (P800) or 225 nm (P225) thickness measured at 35 °C (blue), 50 °C (red) or 70 °C (green).

As expected each additional layer reduces the *WVTR*. At 35 °C the best performing trilayer system made of 20 nm a-C:H, 800 nm parylene and another 20 nm a-C:H (DLC-P800-DLC) reduces the *WVTR* by a factor of ten compared to an uncoated foil (PI) while increasing its thickness by only 1/30. At higher temperatures the effect of the coating is less pronounced with a reduction in *WVTR* by a factor of six at 70 °C. This is probably due to the development of cracks because of the different thermal expansion coefficients of parylene ($35 \cdot 10^{-6}$/K [31]), polyimide ($20 \cdot 10^{-6}$/K according to supplier) and a-C:H ($1 \cdot 10^{-6}$/K to $7 \cdot 10^{-6}$/K [32]).

The calculated *WVTR* of the coating systems without polyimide are represented in **Figure 8** in comparison to nominal values of commercially available parylene C monolayers. Thin (20 nm) hydrogenated amorphous carbon monolayers have a much lower *WVTR* than the thicker parylene monolayers. Adding two a-C:H layers to a parylene layer leads to a reduction in *WVTR* by a factor of thirteen while increasing its thickness by only 5 % (P800) or 18 % (P225).

3.3. In-vivo experience

The presented a-c:H coated stents are on the market since four years. Over 15,000 ureteral stents were sold and to date no problems neither from the patients and urologists nor from the quality management system occurred. Patients as well as doctors report an increased quality-of-life and a facilitated placement procedure.

The complication-free indwelling time is considerably longer due to less frequent painful body implant interactions and significantly reduced bacterial biofilm formation.

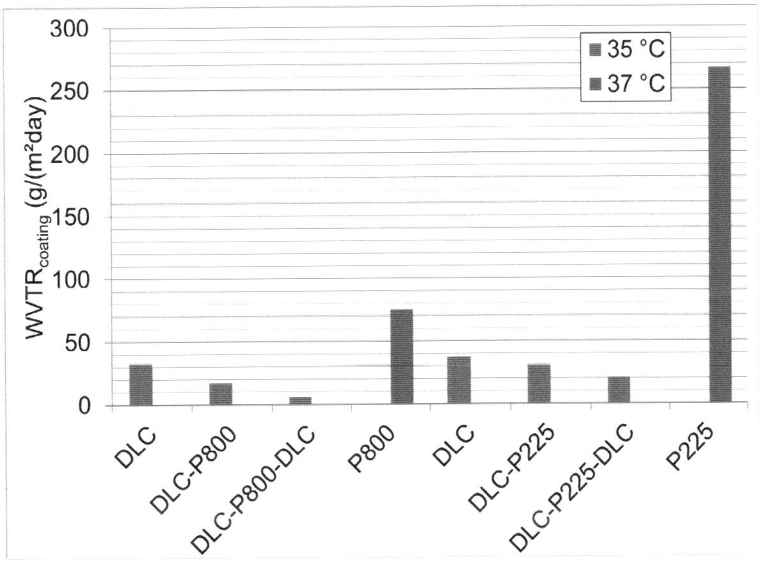

Figure 8: Measured *WVTR* of layer systems of hydrogenated amorphous carbon (DLC) and parylene with 800 nm (P800) or 225 nm (P225) thickness measured at 35 °C (blue) compared to the nominal values of commercially available parylene C monolayers (red) [31].

4. Conclusions

The in-vitro investigations on differently composed a-C:H-films clearly revealed that a stent's encrustation tendency can be stepwise reduced. The especially developed a-C:H single-layer coatings create a durable and highly flexible bidirectional seal on ureteral stents, which has clinically proven to increase the patients' quality-of-life as well as to decrease the formation of crystalline bacterial biofilms, thereby significantly prolonging complication-less indwelling times.

We have presented a custom-built device to analyse the gas diffusion through coated foils in a temperature and humidity controlled environment. Thanks to the detection with a mass spectrometer the same setup can be used to measure the diffusion of various gases in relatively short measurement times. Using standards the obtained values can be compared to other measurement devices. In case of water diffusion the detection limit in transmission rate is below 1 g/day/m^2. Due to the omnipresent water contamination in the vacuum system this value is relatively high. For other gases considerably lower detection limits can be reached.

The investigated a-C:H–parylene–a-C:H trilayer systems reduce water transmission through polyimide foils much better than parylene monolayers of comparable thickness. A trilayer system of 840 nm total thickness, for example, can reduce water transmission by a factor of ten compared to an uncoated polyimide foil. Both types of coatings are deposited by standardised procedures in industrial scale coating devices at substrate temperatures below 40 °C. Therefore the coating systems can easily be applied to a wide range of substrate materials.

The high variability in composition and related physical properties make low-temperature plasma-deposited a-C:H-films a promising concept for the improvement of implants of various bulk materials.

Showing at the same time biofilm and diffusion reducing qualities the potential of a-C:H-coatings may be superior to most other concepts of implant surface modification.

5. Outlook

No "perfect" surface coating will exist that solves all of the problems at the same time that may occur during the interaction between an implant and its surrounding body fluids. Developments towards more individualized solutions may be a strategy to reduce implant-related discomfort, inflammations, infections and more severe medical conditions.

Depending on deposition conditions, density and composition of a-C:H films can be varied in order to provide tailor-made surface solutions required for the particular environmental situation (i.e. the patient's individual medical condition) or desired (active or passive) interaction. Using anchor-sites (e.g. amino-groups, $-NH_2$) the a-C:H coatings can be given specialized functions by controlled covalent binding of 'therapeutically active' molecules (e.g. topically effective drugs, proteins, DNA-fragments, bactericidal agents). More than three amino-groups per nm^2 can be created on a-C:H.

With regard to intra-luminal formation of bacterial and crystalline deposits, in particular at urethral catheters, liquor shunts or indwelling venous cannulae, coating of the implant's inner surface with a-C:H may further rise the implant's complication-free indwelling time. Development of inner a-C:H coatings for tubes with high aspect ratios is in progress.

References

[1] Diamandopolous A 1999 *Nephrol. Dial. Transplant.* **14** 2-9
[2] Stamm WE 1991 *Am. J. Med.* **91** 65S-71S
[3] Warren JW 1992 *Clin. Geriatr. Med.* **8** 805
[4] Warren JW, Muncie HL Jr, Hebel JR and Hall-Craggs M 1994 *J. Am. Geriatr. Soc.* **42** 1286
[5] Kunin CM, Douthitt S, Dancing J, Anderson J and Moeschberger M 1992 *Am. J. Epidemiol.* **135** 291
[6] Yoshikawa TT, Nicolle LE and Norman DC 1996 *J. Am. Geriatr. Soc.* **44** 1235
[7] Lange D and Chew BH 2009 *Ther. Adv. Urol.* **3** 143
[8] Venkatesan N, Shroff S, Jayachandran K and Doble M 2010 *J. Endourol.* **24** 191
[9] Al-Aown A, Kyriazis I, Kallidonis P, Kraniotis P, Rigopoulos C, Karnabatidis D, Petsas T and Liatsikos E 2010 *Ther. Adv. Urol.* **2** 85
[10] Jarvis WR 1996 *Infect. Control Hosp. Epidemiol.* **17** 552
[11] Saint S 2000 *Am. J. Infect. Control.* **28** 68
[12] Cui F and Li D J 2000 *Surf. Coat. Tech.* **131** 481–7
[13] Weiler M, Lang K, Li E and Robertson J 1998 *Appl. Phys. Lett.* **72** 1314
[14] www.ccrtechnology.de/products.php
[15] Morrison NA, Muhl S, Rodil SE, Ferrari AC, Nesládek M, Milne WI and Robertson J 1999 *Phys. Status Solidi A* **172** 79–90
[16] Robertson J 2002 *Mater. Sci. Eng. Rep.* **37** 129–281
[17] Kleinen L, Böde U and Laube N, *Diam. Rel. Mat.* **2008**, *17*, 1746
[18] Angus JC, Koidl P and Domitz S, Carbon Thin Films, in: Plama Deposited Thin Film (Eds.: Mort J and Jansen F), CRC press, Cleveland (1986).
[19] Gorham WF 1966 *J. Polym. Sci. A1* **4** 3027–39
[20] Kavanagh JP 2006 *Urol. Res.* **34** 139
[21] Kleinen L, Böde U, Schenk K, Busch H, Bradenahl J, Müller SC, Hillebrands B and Laube N 2007 *Plasma Process. Polym.* **4** 386
[22] Griffith DP, Musher DM and Itin C 1976 *Invest. Urol.* **13** 346
[23] Grases F, Söhnel O, Villacampa AI and March JG 1996 *Clin Chim Acta.* **244** 45-67
[24] Cox AJ and Hukins DWL 1989 *J Urol* **142** 1347-1350
[25] Gepi-Atee S and Feneley RC 1997 *Proc Inst Mech Eng [H]* **211** 475-478
[26] Busscher HJ, White DJ, Kamminga-Rasker HJ and van der Mei HC 2004 *J. Cryst. Growth* **261**, 87
[27] Wu W and Nancollas GH 1997 *Adv. Dent. Res.* **11** 566

[28] Rosmaninho R and Melo LF 2006 *Int. Diary J.* **16** 81
[29] Quirynen M and Bollen CM 1995 *J Clin Periodontol.* **22** 1-14.
[30] Owens DJ and Wendt RC 1969 *J. Appl. Polym. Sci.* **13** 1741
[31] *Properties of Parylene* (Rosenheim, Germany: Plasma Parylene Systems GmbH) http://www.plasmaparylene.de/medien/medienpool/Eigenschaften-Parylene-2011.pdf
[32] Marques FC, Lacerda RG, Champi A, Stolojan V, Cox DC and Silva SRP 2003 *Appl. Phys. Lett.* **83** 3099

Preparation of nanocomposite carbonaceous films containing palladium nanograins and studies of the annealing time influence on their structure and sensing properties

A Kamińska[1], M Kozłowski[1], P Dłużewski[2], E Czerwosz[1]
[1]Tele and Radio Research Institute, Ratuszowa 11, 03-450 Warsaw, Poland
[2]Instiitute of Physics, Polish Academy of Science, Al. Lotników 32/46,02-668 Warsaw, Poland

E-mail: elzbieta.czerwosz@itr.org.pl

Abstract. C-Pd films were obtained by Physical Vapor Deposition (PVD) or by annealing of these initial PVD films at 650°C in different time (5, 10 and 30 minutes) in argon flow. These C-Pd films were characterized by Scanning Electron Microscopy (SEM), Transmission Electron Microscopy (TEM) and FTIR spectroscopy. The sensing properties of the films were studied in a specially prepared chamber allowing for measurements of the changes in resistivity as a function of gas composition changes. It was found that annealing changes the films morphology, topography and structure, and consequently their sensing properties.

1. Introduction

Since Seiyama [1] developed the first solid-state gas sensor, many other research groups have been attempting to improve gas sensors, with the aim of enhancing their sensing performance. Many materials such as: carbon nanotubes, nanoporous carbon, diamond-like carbon, graphene can detect low concentrations, of ppb order, of gases such as NO_2, NH_3, O_2, H_2, CO_2, and CO [2–8]. However, there are carbonaceous materials with very high resistivity 10^7–10^9 Ω·cm [9,10] that can be applied for sensing different gases by such resistivity limit their sensitivity. Introduction of nanoparticles or nanograins of some metals improves sensing properties. For instance, introduction of palladium nanoparticles into a carbonaceous film allows obtaining a very sensitive hydrogen sensor [11-14]. This phenomenon is connected to solubility of hydrogen in palladium lattice [15-17]. It is known that nanostructural Pd materials improve H_2 sensing in comparison with metallic palladium. In the presence of H_2 the resistivity of Pd will change due to the formation of a solid solution of Pd/H [18]. Furthermore, Pd is highly selective to H_2. Several fundamental problems are associated with bulk Pd-based hydrogen sensors. First, the diffusion of the hydrogen into bulk Pd, e.g. thick Pd film, can cause a large internal stress destroying the film irreversibly [19, 20]. Second, at room temperature the hydrogen atoms diffusion in Pd is very slow (the diffusion coefficient is 3.8×10^{-7} cm^2 /s at 298 K [21]) what causes long time of response for hydrogen. Now, palladium nanomaterials development leads to production of new class of H_2 sensors. These sensors could base on single Pd nanowires [18, 22-24] or on thin films containing nanograins of Pd [25-27]. These films can be obtained by different method (atomic layer deposition, physical vapor deposition or chemical vapor deposition). The change of their resistivity due to hydrogen absorption/desorption is a measure of their sensitivity toward hydrogen. They give promising results taking into account response time, sensitivity and ultralow power consumption.

In our previous papers [11-14, 26-29] we have presented the results of structural, morphological and sensing studies of film composed of nanograins of palladium placed in a carbonaceous matrix. These films were prepared on different substrates (Si, porous Si, Al_2O_3 ceramic) by physical vapor deposition (PVD) method. The changes of the film structure and composition could be controlled by technological parameters. We can also control the size and distribution of Pd nanograins, Pd content in the film and film resistivity. These films exhibited high sensitivity to hydrogen, fast response and selectivity for hydrogen sensing. In this paper, we present the results of studies of the influence of annealing of these films on their sensing properties. Films obtained by PVD method were annealed at 650°C in an argon atmosphere for different time. We found that annealing of non-conductive C-Pd films could lead to the film structure changes causing change of its conductivity. Conductive paths based on palladium nanograins are formed in these C-Pd films. The sensitivity of these films to hydrogen was also discussed.

2. Experimental

The nanostructural C-Pd films were prepared by PVD (Physical Vapor Deposition) from two separate sources containing: fullerene (first) and palladium acetate (second). In this process, performed in dynamic vacuum with the pressure of 10^{-6} mbar, films were obtained by deposition of fullerene C_{60} (Sigma-Aldrich, 99.9%) and palladium acetate $Pd(OAc)_2$ (Sigma-Aldrich, 99.98%). These films were deposited on Al_2O_3 substrates, with the deposition time t = 10 min. The roughness of the substrate was 0.13-0.2 μm. Currents through both sources were the same for all prepared films. The distance - d between the sources and the substrate was changed and it was 54 mm for sample S1 and 69 mm for S2. This caused a small difference of temperatures on the substrates: T=59°C for S1 and T=54°C for S2 and bigger difference in Pd content (see Tab.1). The temperature of substrate was measured during all PVD process and given here temperature is the final temperature registered and the final moment of this process. It should be noticed that temperature for both PVD processes (for sample S1 and S2) stabilized after 4-5 min from the beginning of the process. The temperature was measured with thermocouple fixed always at the same place of the sample. These measurements were repeated several times to estimate the errors connected to the method of thermocouple mounting and heating time of sources. Films prepared by PVD were annealed at 650°C. The annealing was performed in argon atmosphere (to prevent oxidation of palladium), in three different time values: 5, 10 and 30 minutes. The films annealed at different time are marked as S1A(5/10/30) and S2A(5/10/30).

The palladium content in the initial films obtained by PVD was determined by atomic absorption spectra for a proper solutions of films prepared by wet chemical analysis.

The resistivity of PVD initial films was measured in an experimental set-up, prepared for testing a gas influence on electrical properties of films (Figure 1). The changes of the electrical resistivity of C-Pd films at 1%H_2/N_2 mixture of gases were measured under room conditions (temperature and pressure). The flow rate of 1%H_2/N_2 gaseous mixture was kept constant and it was 50 ml/min. After absorption of hydrogen, the air was introduced (with a rate of 50 ml/s) to the experimental chamber to remove H_2 gas. Several hydrogen absorption-desorption cycles were carried out for each sample. After desorption, the resistance reached the final value after 30 minutes and remained stable afterwards. It should be mentioned that obtained in this way the response of the film for hydrogen cannot be used for determination of real response time, that is the immediate response of film to contact with gas. In this experiment we register time of total exchange of gas and air in experimental chamber.

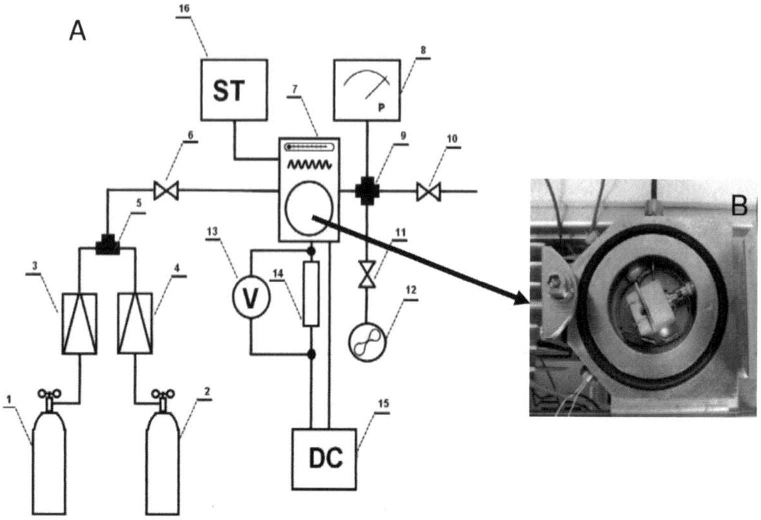

Figure 1 A) Experimental setup for sensing measurements (1, 2 – gas bottles, 3, 4 – mass flow controllers, 5 – gas mixer, 6, 10, 11 – valves, 7 – measurement chamber, 8 – pressure meter, 9 – four-way splitter, 12 – vacuum pump, 13 – voltmeter, 14 – reference resistor, 15 – voltage source, 16 – temperature controller). B) Photo of the measurement chamber

Technological parameters, concentration of Pd and resistivity of initial PVD films and annealed films are gathered in Table 1.

Table 1. Technological parameters, concentration of Pd and resistivity for C-Pd films

Sample number	*d* [mm]	Pd content [wt %]	Annealing time [min]	Resistivity [10^6 Ω·cm]
S1	54	17		>60
S2	69	10		>60
S1A5			5	>60
S1A10			10	>60
S1A30			30	>60
S2A5			5	29.9
S2A10			10	21.9
S2A30			30	6.1

The films were also characterized by Scanning Electron Microscope (SEM) and Transmission Electron Microscope (TEM). SEM investigations were performed with the use of JEOL JSM-7600F electron microscope with a Secondary Electron (SE) detector and with a Low Angle Backscattered Electron (LABE) detector. LABE mode is used to present a contrast composition of the film components (C and Pd). The field emission SEM was operated at 5 keV incident energy. TEM investigations and electron diffraction from a selected area (SAED) were performed for some selected samples. These investigations were performed with JEOL JEM-2000EX electron transmission microscope operating at 200keV incident electron beam energy.

FTIR spectra were obtained with ThermoScientific Nicolet iS10 FTIR spectrometer, using Attenuated Total Reflectance (ATR) in the spectral range of 650-4000 cm^{-1}, at the spectral resolution of 4 cm^{-1}.

3. Results and discussion

3.1.PVD C-Pd films

Neither S1 nor S2 films were sensitive to hydrogen, while analyzing their resistivity changes in a hydrogen presence.

Initial films prepared by PVD method contain 10 and 17 wt. % of Pd. The resistivity of these films was very high (>$6x10^7$ $\Omega\cdot$cm) and did not change when the films were introduced into the chamber filled with a H_2/N_2 mixture. Such high resistivity could be attributed to palladium nanograins distribution in a carbonaceous matrix. This is shown in TEM images for samples S1 and S2. In Fig. 2a and Fig. 3a TEM images of S2 and S1 samples are presented, respectively. Pd nanograins are seen as dark objects. These Pd nanograins are rarely distributed and they cannot form conduction paths, even when PdH_x grains are formed. In TEM image presented Fig. 2d one can see the area with fullerene molecules forming fullerite with *fcc* type structure. In this image the distance d_{111} for this structure is marked. In Fig. 2b and Fig. 3b selected area electron diffraction (SAED) patterns from the fragments of films are shown. The SAED pattern for the sample S1 consists of diffuse rings with the low diameters, related to diffraction on very small amorphous carbonaceous grains and diffraction rings, and with larg diameters attributed to very small Pd nanograins. The Pd nanograins size distribution analysis for sample S1 shows that the diameter of nanograins is between 4 and 16 nm (see diameter size distribution histogram in Fig. 3c). The SAED pattern for sample S2 is different and contains diffuse rings that could be attributed to diffraction on C_{60} fullerite nanograins (internal rings) and reflexes related to diffraction on large Pd grains. These nanograins sizes can be determined from size distribution histogram and they are 5-100 nm.

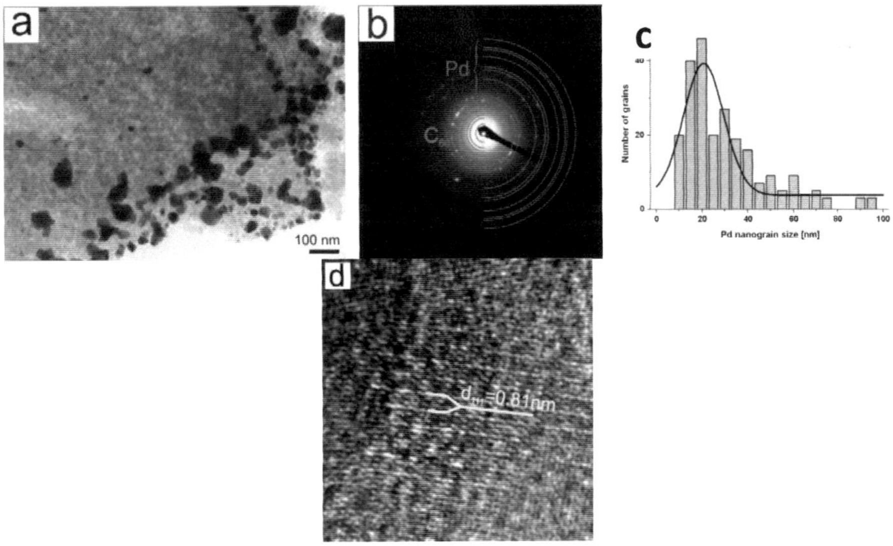

Figure 2 a) TEM image for sample S2; b) SAED pattern for S2; c) Pd nanograins size distribution for S2, d) TEM image of the area with fullerite structure

Journal of Physics: Conference Series **564** (2014) 012002 doi:10.1088/1742-6596/564/1/012002

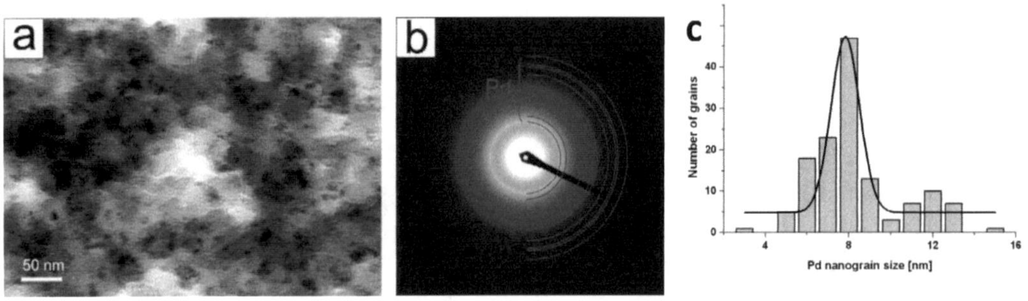

Figure 3 a) TEM image for sample S1; b) SAED pattern for S1; c) Pd nanograins size distribution for S1

S1 and S2 films have similar molecular structure, which is confirmed by FTIR spectra (Fig. 4). Bands marked with arrows can be attributed to the vibrations of C_{60} molecule and C-O groups in palladium acetate. For a comparison spectra of fullerene and palladium acetate are also presented. Based on these spectra we can deduce that palladium acetate molecules or clusters of molecules are present in films. It should be mentioned that the carbonaceous matrix is non-conductive when is based on fullerenes molecules. Decomposition temperature of Pd acetate is much lower (210-290 ^{O}C) than final temperature of source (800-1100 ^{O}C) containing Pd acetate. For some experiments (with low heating ratio), the temperature of the source containing acetate is low during the first moment. Then, palladium acetate could not be decomposed completely and this could be a reason why in FTIR spectra we can observe a trace of Pd acetate. When the heating ratio is high enough no trace of acetate should be observed.

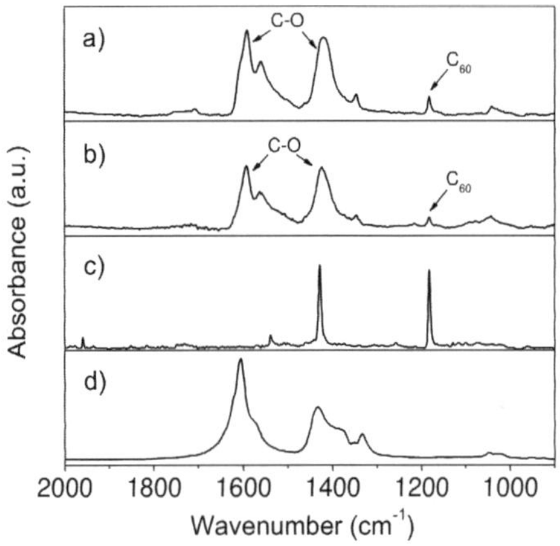

Figure 4 FTIR spectra of a) S1, b) S2, c) fullerene C_{60} and d) palladium acetate

SEM images of samples S1 and S2 are similar and they show that the film topography is dominated by the substrate surface shape (big grains of one or a few μm in size). Oval grains with the diameter of ~200-300 nm can be observed on these grains. The example of a typical topography of these samples is presented in Fig. 5a and b. We did not observe Pd nanograins even at the highest resolution of SEM images. Non-conductive character of samples made difficult obtaining an sharp image at the highest magnification.

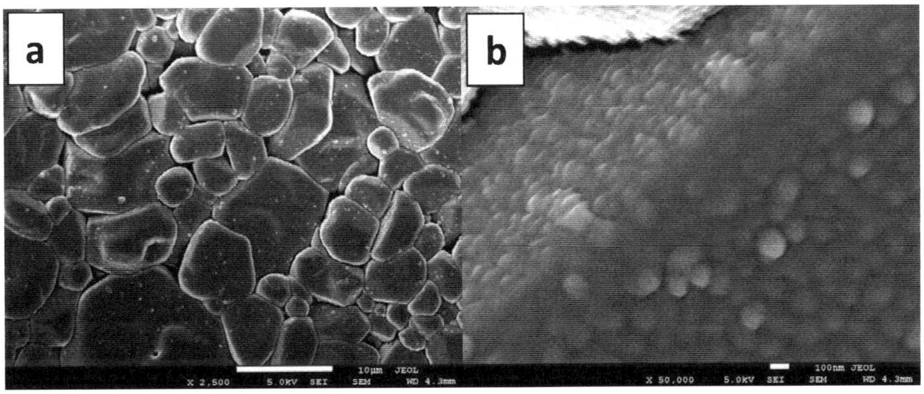

Figure 5 SEM image of initial PVD C-Pd film: a) magnification 2500; b) 50000

3.2 Annealed films

S1 and S2 film after annealing under the same conditions reacted with hydrogen in a different way. S1 film annealed for 5, 10 and 30 minutes did not change its resistivity in the presence of hydrogen. S2 annealed samples changed their resistivity (decrease) and they were highly sensitive to hydrogen. In Fig.6 are presented the resistivity changes due to interaction with a 1% H_2/N_2 gas mixture.

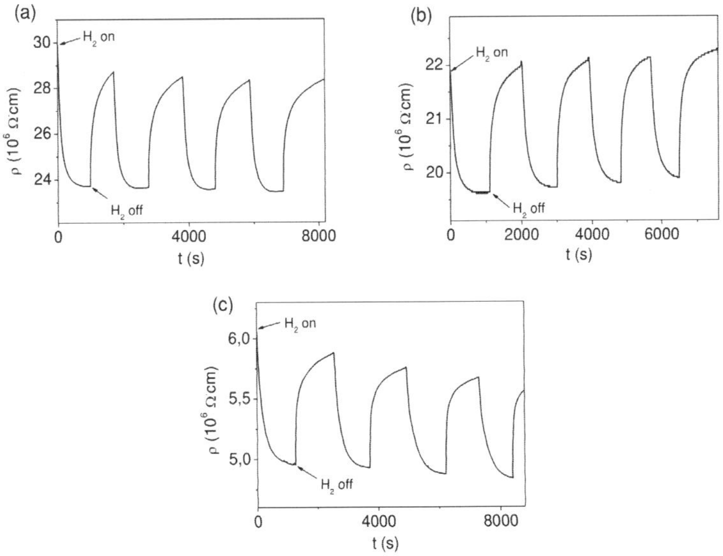

Figure 6 Changes of resistivity in 1% H_2/N_2 for the following samples: a) S2A5, b) S2A10 and c) S2A30

SEM and TEM images of samples S1A5 and S1A10 are shown in Fig. 7. The grains of palladium are of different size (up to a few hundreds nm) and they are randomly distributed on the sample surface. The agglomeration of small palladium nanograins increases with the annealing temperature and for S1A30 sample leads to the growth of very long Pd grain (a few μm). Most of Pd grains are small and the carbonaceous matrix seems tight. SAED patterns prove an amorphous character of the matrix. These amorphization of carbonaceous matrix is due to catalytic reaction of palladium with this matrix undergoing in high temperature. These diffraction patterns also exhibit reflexes related to the Pd *fcc* crystalline type structure (Fig.7d).

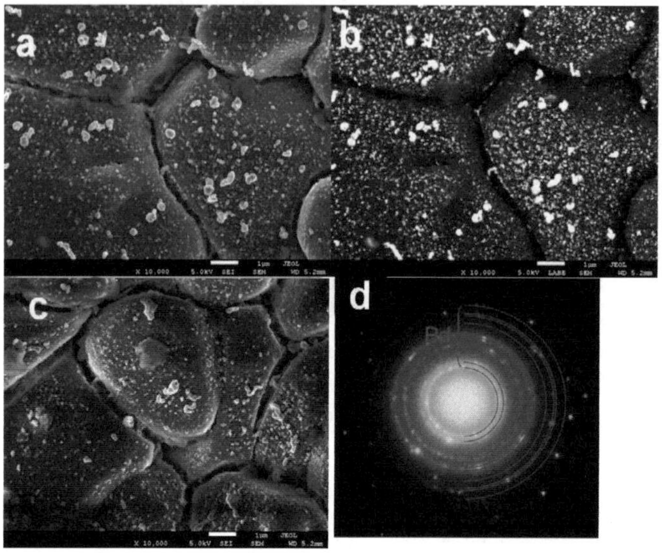

Figure 7 SEM images of samples S1A5 and S1A10: a) SE image and b) LABE image for S1A5, c) SE image for S1A10 and d) SAED pattern from S1 annealed film. Bright objects in SEM images are palladium grains.

Fig.8 presents SEM images of S2 film annealed for 5, 10 and 30 minutes. For all theses samples we observed the effect of formation of conductive paths composed of palladium grains collected on the Al_2O_3 grains border. For samples S2A5 and S2A10 the form of such a conductive path is similar and Pd grains packing along the path is similar. For sample S2A30 it seems that Pd grains do not form the paths, but they are uniformly distributed in all volume of the film. The carbonaceous matrix has a porous form for all these samples. More pores are formed as the annealing time increases (Fig.9a-c). The diameter of the pores decreases with the increasing annealing time. Palladium grains are placed in these pores and form a very dense network of connections. Palladium grains placed in the pores are the smallest in sample S2A30. This effect could cause lowering of the resistivity of this sample. During the films interaction with hydrogen the volume of palladium grains increases and the network of these grains becomes more effective, because of a closer contact between them. This effect could have a stronger impact to the film conductivity than the effect of PdH_x formation, which reduces the film conductivity due to PdH_x resistivity being higher than the resistivity of palladium. This mechanism could be responsible for the observed decrease in the film resistivity in the presence of hydrogen.

The film porosity could also affect its resistivity. It was demonstrated [30] that the resistivity of porous carbon material is ~10^2 – 10^3 $\Omega \cdot$cm and decreases as the temperature goes up. The combination of these two above mentioned effects could also diminish resistivity of film. However the porous carbonaceous structure could play a role in hydrogen absorption and hydrogen-carbon stable bonds could be formed, which results in lowering the baseline during hydrogen sensing measurements (Fig. 6c). Such resistivity changes during sensing measurements are connected with non-total desorption of hydrogen from the film.

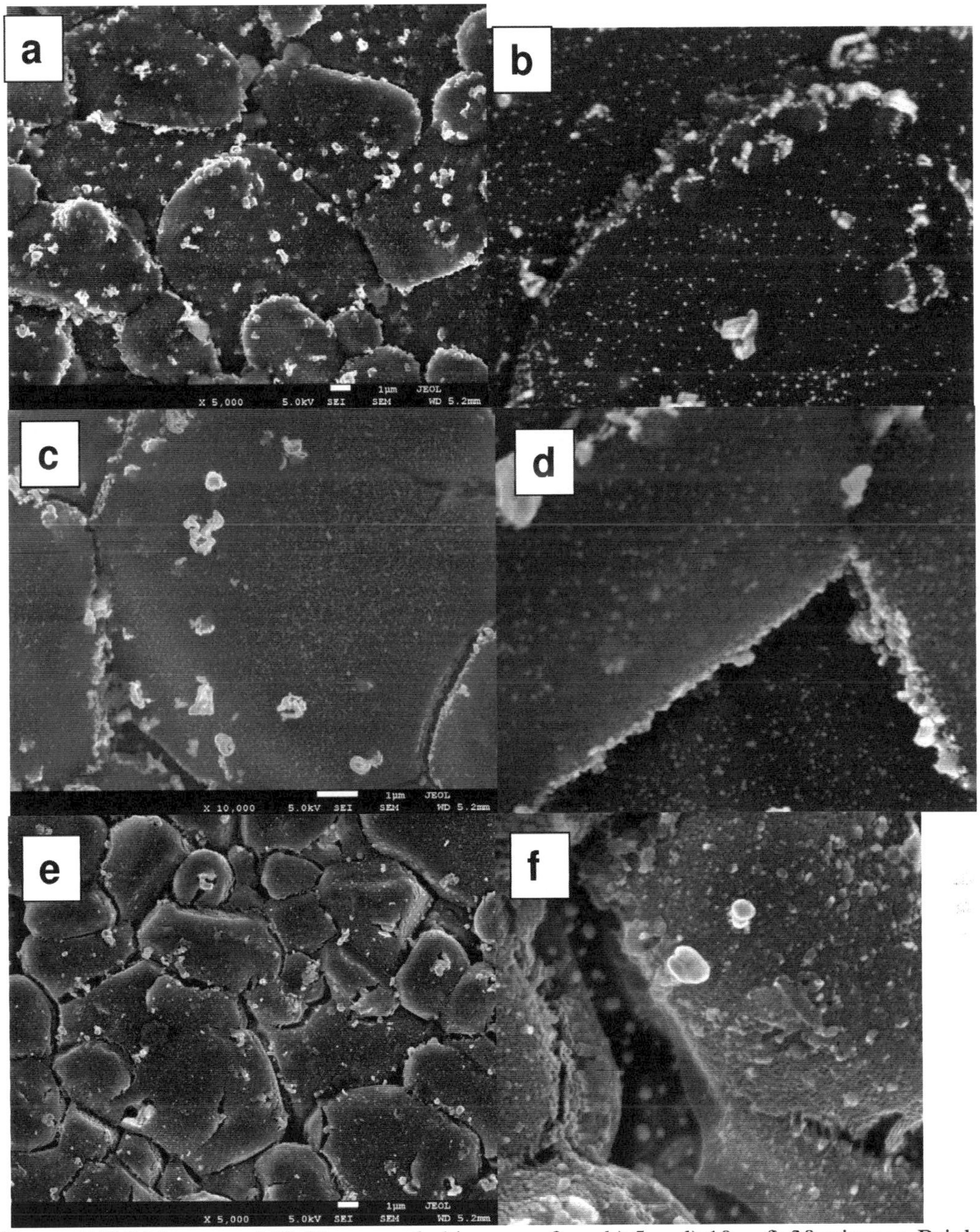

Figure 8 SEM images of S2 film annealed in argon for a,b) 5, c,d) 10, e,f) 30 minutes. Bright objects are palladium grains..

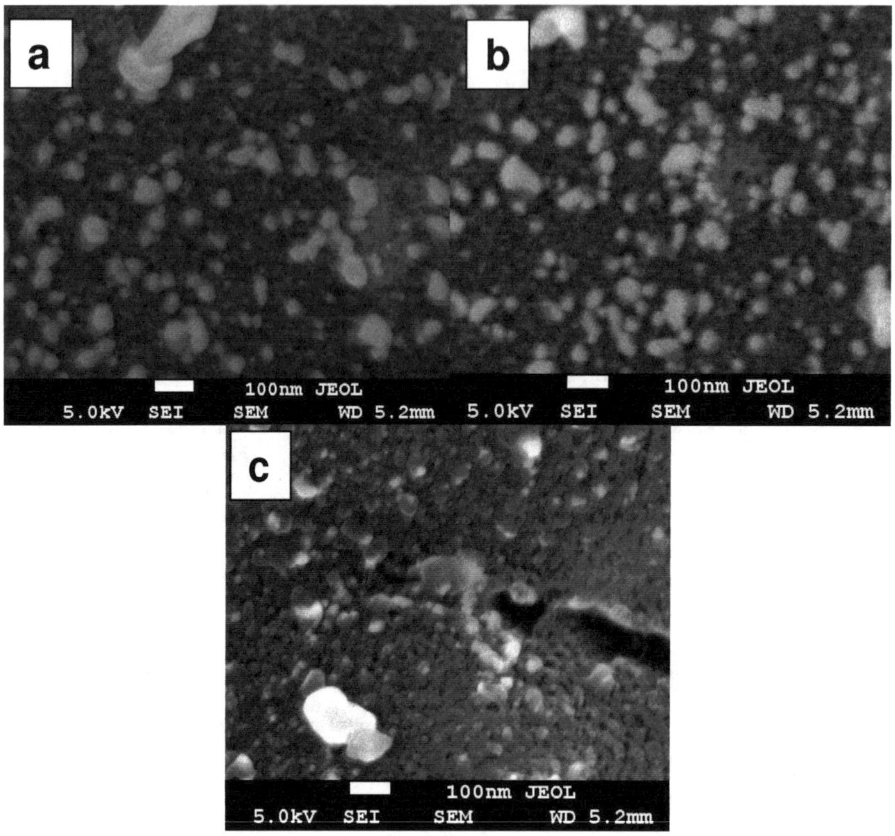

Figure 9 SEM images of pores in S2 film annealed in argon for a) 5, b) 10 and c) 30 minutes.

4. Conclusions

In conclusion, annealing of C-Pd films in argon causes an improvement in the sensing properties for films with 10 wt% of Pd. Annealing of such non-conductive C-Pd films obtained by PVD method could improve their conductivity due to a formation of Pd nanograins conductive paths and a porous carbonaceous matrix. Annealing allows for such modification of the structure of C-Pd films that leads to the formation of palladium conductivity paths and formation of a network of palladium nanograins placed in the pores of the carbonaceous matrix.

The film containing less Pd seems most suitable for application in hydrogen sensors. Annealing of PVD films at different time improves their initial sensing properties (response time lower than 100 s, low noise to signal ratio, high response value).

5. Acknowledgements

This project is co-financed by the European Regional Development Fund within the Innovative Economy Operational Programme 2007-2013 No. UDA-POIG.01.03.01-14-071/08-09.

References

[1] Seiyama T, Kato A, Fujiishi K and Nagatani M,1962 *Anal. Chem.* **34** 1502
[2] Collins P G, Bradley K, Ishigami M and Zettl A, 2000 *Science* **287** 1801

[3] Kong J, Chapline M G and Dai H, 2001 *Adv. Mater.*, **13** 1384
[4] Sotiropoulou S and Chaniotakis N A 2005 *Analytica Chimica Acta* **530** 199
[5] Chih-Cheng Lu, Kuan-Hsun Liao, Jeff Tsung-Hui Tsai, Chueh-Yang Liu and Jih-Perng Leu, 2009 *Biomed. Eng. Appl. Basis Commun.* **21** 399
[6] Luo J K, Fu Y Q, Le H R, Williams J A, Spearing S M and Milne W I, J. 2007 *J.Micromech. Microeng.*, **17** S147
[7] Geim A K 2009 *Science* **324** 1530
[8] Schedin F, Geim A K, Morozov S V, Jiang D, Hill E H, Blake P and Novoselov K S 2007 *Nat. Mater.* **6** 652
[9] Bruzzi M, Piseri P, Miglio S, Bongiorno G, Barborini E, Ducati C, Robertson J and P. Milani, 2003 *Eur. Phys. J.*, **B 36** 3
[10] Rode A V, Gamaly E G and Luther-Davies B, 2000 *Appl. Phys,.* **A 70** 135
[11] Kowalska E, Czerwosz E, Kozłowski M, Surga W, Radomska J and Wronka H, 2010 *J. Therm. Anal. Calorim.*, **101** 737
[12] Czerwosz E, Dłużewski P, Kowalska E, Kozłowski M and Rymarczyk J, 2011 *Physica Status Solidi (c)*, **8** 2527
[13] Molenda K, Kamińska A, Krawczyk S, Kozłowski M, Czerwosz E and Wronka H, 2011 *Proc. SPIE* **8008**, 800821
[14] Kowalska E, Czerwosz E, Kamińska A and Kozłowski M, 2012 *J. Therm. Anal. Calorim.*, **108** 1017
[15] Lacher J R, 1937 *Proc. R. Soc. Lond.,* **161** 525
[16] Sieverts A, Bruning H, 1932 *Z.Phys.Chem.*, A, **163,** 409
[17] Sieverts A, Hagen H, 1935 *Z.Phys.Chem.*, A, **174**, 247
[18] Offermans P; Tong H D; van Rijn C J M; Merken P; Brongersma S H; Crego-Calama M, 2009 *Appl. Phys. Lett.,* **94** 223110
[19] Pundt A, 2004 Adv. Eng. Mater., **6** 11
[20] Lee E, Lee J M, Koo J H, Lee W, Lee T, *2010 Int. J. Hydrogen Energy* **35** 6984
[21] Xu T, Zach M P, Xiao Z L, Rosenmann D, Welp U, Kwok W K, Crabtree G W, 2005 *Appl. Phys. Lett.*, **86** 203104
[22] Jeon K. J, Jeun M, Lee E, Lee J M, Lee K I, von Allmen P, Lee W, 2009 *Nanotechnology* **19** 495501
[23] Jeon K J, Lee J M, Lee L., Lee W, 2009 *Nanotechnology* **20** 135502
[24] Yang F, Taggart D K, Penner R M, 2010 *Small* **6** 1422
[25] Elam J W, Zinovev A, Han C Y, Wang H H, Welp U , Hryn J N, Pellin M J, 2006 *Thin Solid Films* **515** 1664
[26] Kozłowski M, Diduszko R, Olszewska K, Wronka H, Czerwosz E, 2008 *Vacuum* **82** 956
[27] Kowalska E, Czerwosz E, Diduszko R, Kamińska A, Danila M, 2013 *Sensors and Actuators* **A 203** 434
[28] Kamińska A, Krawczyk S, Czerwosz E, Sobczak K, Kozłowski M, 2013 *Sensors and Actuators* **A 196** 86
[29] Kamińska A, Krawczyk S, Kozłowski M, Czerwosz E, Sobczak K, 2013 *Sensor Letters* **11** 500
[30] Kennedy L J, Vijaya J J, Sekaran G, 2005 *Mater. Chem. Phys.* **91** 471

Influence of CVD process duration on morphology, structure and sensing properties of carbonaceous-palladium films

E Kowalska[1], A Kamińska[1], E Czerwosz[1], M Kozłowski[1], K Sobczak[2], B Witkowski[2]
[1]Tele & Radio Research Institute, Ratuszowa 11 Street, 03-450 Warsaw, Poland
[2] Institute of Physics PAS, al Lotników 32/46, 02-668 Warsaw, Poland

E-mail: ewa.kowalska@itr.org.pl

Abstract. We present the nanocomposite carbonaceous-palladium (C-Pd) thin films prepared by physical vapor deposition (PVD) and chemical vapor deposition (CVD) methods. Scanning electron microscope (SEM) and transmission electron microscope (TEM) methods were used to study the topography, morphology and structure of carbon and palladium nanograins contained in these films. The quantitative analysis of the elemental composition of C-Pd films was determined by energy-dispersive spectroscopy (EDS). The initial PVD films were modified in a CVD quartz reactor using xylene (the mixture of isomers) as a modifying factor at different times (5, 10 and 30 minutes) at a constant temperature of 650°C in atmospheric pressure. It was observed that the average size of palladium nanograins increased with an increasing duration of modification process. The differences in microstructures observed in the CVD films modified at different times, affect their response in measurements of resistance changes in the gas containing H_2 in various concentrations. All samples were measured by cathodoluminescence (CL) method. In CL studies a large amount of objects with high intensity of CL was found. Some of them show the emission bands both at 450 nm and 750 nm. Other reveals emission band only at 450 nm. CL observations show that Pd nanograins coated by graphite shells exhibit optical activity.

1. Introduction

Nanostructured materials containing different allotropic form of carbon such as foam, graphite, fullerene C_{60}, nanotubes, nanofibers doped with transition metals e.g. palladium, platinum, nickel, iron, and cooper can be used in various applications e.g. in catalysis, as hydrogen storage materials [1-4], as hydrogen sensors [5-7] or can provide electrodes in batteries [8], fuel cell [9] or can be used as electrochemical biosensors [10]. These materials are characterized not only by highly developed surface area of carbon but also by presence of metals nanograins which play a special role in catalysis. These metals nanograins cause decomposition of chemical compounds adsorbed on them.

Nanostructures based on nanoporous carbon matrix containing palladium nanograins called as the C-Pd films are promising materials for hydrogen sensing applications. Presently hydrogen technologies awaken of great interest among scientists and entrepreneurs. This interest is related to the potential use of hydrogen as an energy carrier. Recently, the growth of the hydrogen importance in the world economy e.g. in transport, chemical compounds production, cryogenic systems, electronic and metallurgical industry has been noticeable.

[1] ewa.kowalska@itr.org.pl

Properties such as topography, morphology, electrical and hydrogen sensing of C-Pd films prepared by us, were described in our previous papers [11-17]. One of the factors which decide about using such materials for H_2 detection is presence of Pd nanograins embedded in carbon matrix. On the other hand a well-developed specific surface area of carbonaceous matrix, containing pores with different size and distribution could also play a positive role in hydrogen sorption. Palladium is one of the elements which are capable of reversibly absorbing large amounts of hydrogen [18]. Molecular hydrogen is dissociated at the palladium surface into two H atoms before absorption. Two H atoms can recombine to H_2 in the desorption process. The thermodynamic aspects of hydride formation from gaseous hydrogen are described by pressure–composition isotherms which were presented in papers [19, 20]. The host metal initially dissolves some hydrogen as a solid solution (α-phase). As the hydrogen pressure together with the concentration of H in the metal is increased, interactions between hydrogen atoms become locally important, and start to nucleation and growth of the hydride (β-phase). While the two phases (α and β) coexist, the isotherms show a flat plateau, the length of which determines how much H_2 can be stored reversibly with small pressure variations. In the pure β-phase, the H_2 pressure rises steeply with the concentration.

In this paper we present carbonaceous-palladium films obtained by physical vapor deposition (PVD) method and next modified by chemical vapor deposition (CVD) technology as a potential hydrogen sensitive active layer in the hydrogen sensor. These films exhibit promising solid-gas properties toward hydrogen sorption. Such materials combine physisorption at the carbon pore surface and hydrogenation of the metal forming hydride. The aim of the present work is to study the influence of duration of CVD process on topography, morphology and hydrogen sensing properties of C-Pd films. In order to show difference in the carbon structure and palladium nanograins size in films obtained by CVD modification at different times (5, 10 and 30 minutes) we used SEM, EDS, TEM and CL methods.

2. Experimental

2.1 PVD preparation and CVD modification of C-Pd films

Nanocomposite carbonaceous – palladium films were prepared by physical vapor deposition process. These films were deposited on alumina (Al_2O_3) substrates using as films' precursors: fullerite C_{60} powder (99.99%, Sigma-Aldrich) and palladium(II) acetate ($Pd(OAc)_2$ Sigma-Aldrich). Both compounds were evaporated from two separated sources using electrical heaters. PVD process was carried out in a dynamic pressure of 10^{-3} Pa. The current (I) flowing through the fullerene source was 2,1 A and through palladium acetate source was 1,2 A. Duration time (t) was 8 minutes and distance (d) between sources and substrates was 54 mm. As a result C-Pd films with thickness of 300 nm were obtained what suggesting that the growth rate was 37.5 nm/1min. Next PVD films were modified by CVD method using xylene (the mixture of isomers - Chempure) as an additional carbon source. Xylene was used due to easily decomposition in CVD temperature forming different molecular ions (type: $C_3H_3^+$, $C_4H_2^+$, $C_4H_3^+$, $C_nH_n^+$ or $C_nH_{n-1}^+$ [21]), these ions can react with surface of C-Pd film causing an increase its specific surface area. They additionally form graphite shells around Pd nanograins which were confirmed by TEM method. The expanded surface intensified the reaction with hydrogen.

Process was performed in a quartz reactor at temperature of 650°C in an argon flow rate of 40L/h in an atmospheric pressure. Duration time of CVD modification was varied from 5 to 30 minutes. As the CVD modification result, carbon matrix was converted into porous structures and palladium nanograins appeared on films' surface. Sample modified by CVD in 5, 10 and 30 minutes was named as S5, S10 and S30, respectively.

2.2. SEM and EDS characterization

The topography, morphology and structure of PVD initial films and films after CVD modification (S5, S10 and S30) was studied by Scanning Electron Microscopy (SEM). These investigations were

performed with the field emission scanning electron microscope JEOL JSM-7600f operating with 5keV incident energy. SEM studies were carried out using two modes of operation: SE (secondary electron) and LABE (low angle backscattered electron). In this mode, the contrast depends on the atomic number of elements of the observed microstructure. Areas with higher atomic number are mapped as bright, while those of lower atomic number - as darker.
The quantitative analysis of the elemental composition of CVD films was determined by energy-dispersive spectroscopy (EDS). The EDS measurements were performed with INCA ENERGY 250 using accelerating voltage of 7 kV with the area scanning method.

2.3. TEM characterization
Transmission Electron Microscopy (TEM) was applied to study structure of S5, S10 and S30 films. The specimens for TEM were prepared by direct moving the C-Pd films from substrate to Cu 1000 mesh grids. The investigation was carried out on JEM 2000EX microscope operating with 200 keV electron beam energy. Electronograms were recorded at fixed camera 600 mm. Observations were performed using a diffraction contrast using technique of bright field – images are formed from the incident beam. The grain size histograms were obtained by considering a region of a sample having about 250 nanocrystals and approximating the shape of each nanocrystal by a sphere. The obtained histograms were fitted to log-normal distributions.

2.4. Cathodoluminescence measurements
Cathodoluminescence measurements were performed on Scanning Electron Microscope Hitachi SU-70 equipped with cathodoluminescence GATAN MonoCL3 system. In measurements were used accelerating voltage of 15kV, photomultiplier as detector and diffraction grating of 300 lines per inch.

2.5. Hydrogen sensing measurements
The sensor evaluation of CVD films was performed in a special measurement set-up. Conductive silver paste (SPI Supplies) was used to make the electrical contacts to these films. These films were placed in the measurement chamber, having electrical feed through, gas flow inlet and outlet. Mass flow controllers were used to ensure a constant, required gas flow rate of 50 ml/min. Hydrogen sensing measurements were performed for different H_2 concentration in nitrogen: 0.5 vol.%, 1 vol.% and 2 vol.%. For each hydrogen concentration, three H_2 absorption/desorption cycles were performed. The process of hydrogen desorption was carried out in the flow of synthetic air. Measurements were performed at room temperature and under atmospheric pressure.

3. Results and discussion

3.1 SEM and EDS investigations
In figure 1 SEM images in SE mode of PVD films with various magnification are presented. On the film's surface carbonaceous nanograins are visible. These nanograins have different sizes, posses rounded shapes and smooth walls. Pd nanograins are not observed on film's surface. According with FTIR investigations the remnant of fullerene C_{60} and palladium acetate was observed in PVD films [22].

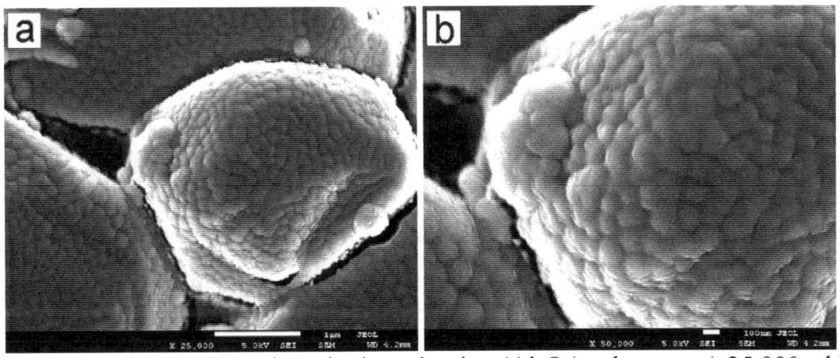

Figure 1 SE images of PVD films deposited on alumina (Al_2O_3) substrate: a) 25 000x, b) 100 000x

After CVD modification, morphology, topography and structure of PVD films are drastically changed. SE and LABE images of S5 film are presented in figure 2(a, c) and 2b, respectively. As can be seen carbon matrix was transformed into the porous structure, containing on the surface palladium nanograins distributed as evenly. By comparing SE and LABE images it can be seen that Pd nanograins are very fine and are surrounded by graphite shells what was confirmed by TEM method [16]. In some fragments of this film carbon matrix is cracked.

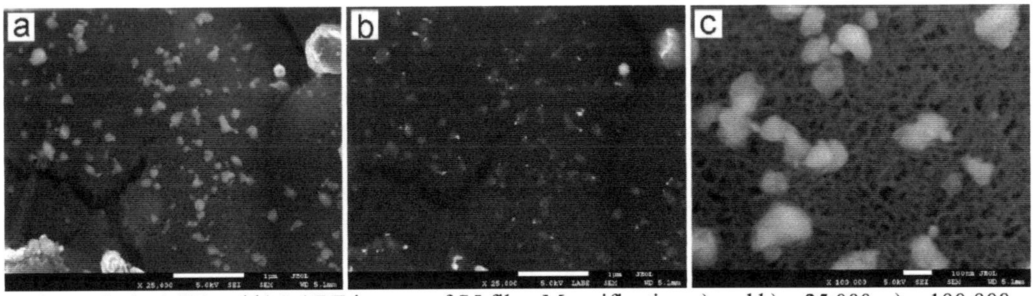

Figure 2 a), c) SE and b) LABE images of S5 film. Magnification a) and b) x 25 000; c) x 100 000

SE and LABE images of S10 film are showed in figure 3(a, c) and 3b, respectively. In this case a greater density distribution of Pd nanograins is observed. Additionally in some areas bigger palladium nanograins are visible. In LABE image around some of Pd nanograins graphite shells are also observed.

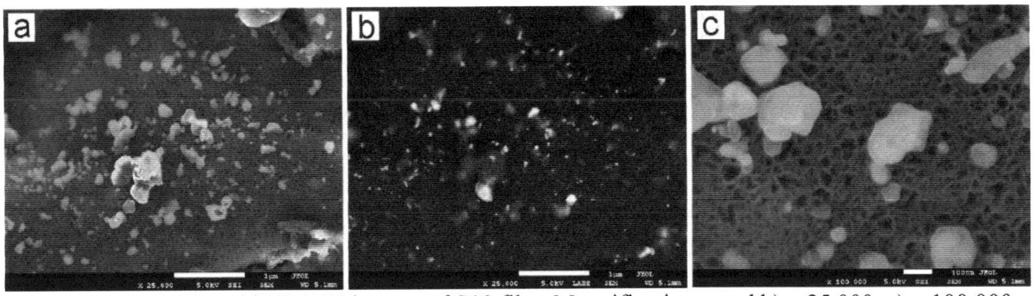

Figure 3 a), c) SE and b) LABE images of S10 film. Magnification a and b) x 25 000; c) x 100 000

In figure 4(a, c) and 4b SE and LABE images of S30 film are showed, respectively. On the film's surface uniformly distributed very fine Pd nanograins are observed. TEM studies show that these fine nanograins have the graphite shells. It should also be noted that the larger grains with a diameter above

300 nm were found at the boundaries of substrate grains. The larger Pd nanograins were also encapsulated in graphite shells.

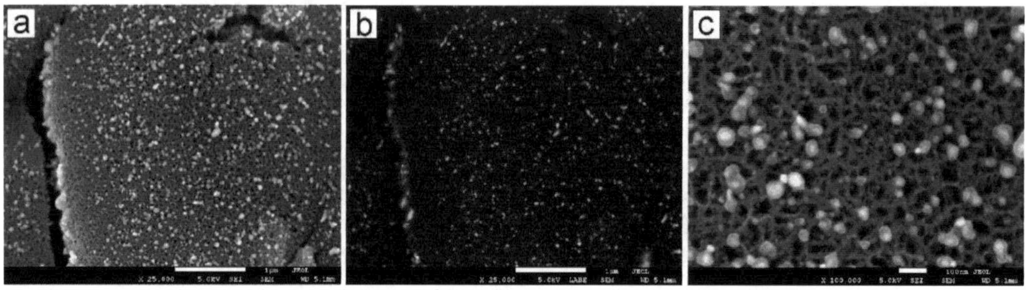

Figure 4 a), c) SE and LABE images of S30 film. Magnification a) and b) x 25 000; c) 100 000

EDS analysis was performed at an accelerating voltage of 10 kV at three different areas. In table 1 the average values of the 3 measurements. Since the films thickness does not exceed 200-300 nm, in all cases the Al and O signals recorded from a substrate Al_2O_3 were registered. It was observed that with an increase in CVD process duration, the films thickness decreased. Therefore, the measured palladium concentration is not a real concentration of this element in all measured films (especially in S30 film which is the thinnest). Thus, the ratio of Pd to C (Pd/C) is more an authoritative indicator comparative. The Pd/C increases with increasing duration of CVD process.

Table 1 EDS analysis of S5, S10 and S30 films

Elements	S5*	S10*	S30*
C	73,64	69,33	65,29
	16,46	16,46	15,02
Al	4,68	7,41	12,23
Pd	5,22	6,77	7,46
Pd/C	0,07	0,09	0,11

* All data are given in weight%

We suggest that reduction of films thicknesses for a longest process is caused by sublimation of remnant of C_{60} and also an evaporation of palladium acetate. Within the first 5 and 10 minutes is not enough time to the fullerene sublimation occurred (the time may be too short). In case of the longer time in TEM studies we observed, that those Pd nanograins migrates to the surface films and they connect to the larger grains. The longer time provoke Pd nanograins migration and formation of larger grains.

3.2 TEM characterization

The electron diffraction (ED) pattern is presented in figure 5. All films after CVD modification in various times show the similar electron diffraction pattern. Analysis of ED allows identifying structure of palladium as *fcc*. In figure 5 the graphite plane {00.2} is also visible. The graphite presence is caused by xylene in CVD process. Our previous studies [16, 17] have showed that annealing of PVD film in argon atmosphere at temperature of 650°C without the xylene presence, did not cause the graphite shells formation. However, in CVD process with the xylene, graphite shells around Pd nanograins were formed.

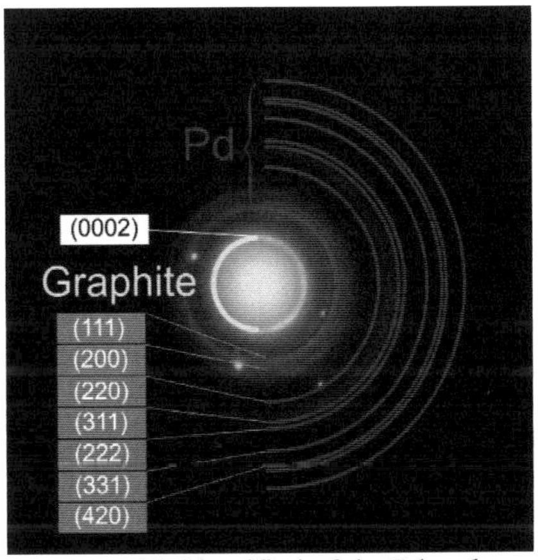

Figure 5 Electron diffraction pattern with identified of the Pd and graphite structure for all S5, S10 and S30 films. The lattice planes of graphite and palladium are indexed

In figures 6, 7 and 8 we present TEM images and histograms of Pd nanograins size distribution for films modified at 5 minutes (S5), 10 minutes (S10) and 30 minutes (S30), respectively. We observe that the average size of Pd nanograins increases with increasing time of the CVD modification. For S5 film the average size of Pd grains is 42 nm, and for S10 film is 80 nm. Longer time does not significantly increase the size of Pd grains, and for S30 film it is 83 nm. Although in figure 8a in TEM image we can see Pd nanograins with a diameter above 300 nm.

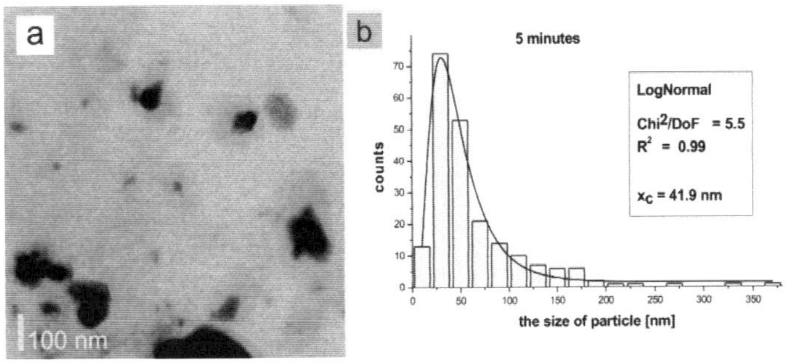

Figure 6 a) TEM image of S5 film and b) histogram of Pd nanograins size distribution.

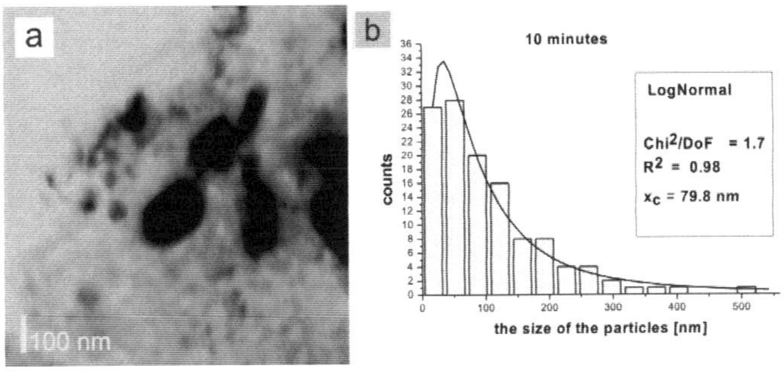

Figure 7 a) TEM image of S10 film and b) histogram of Pd nanograins size distribution.

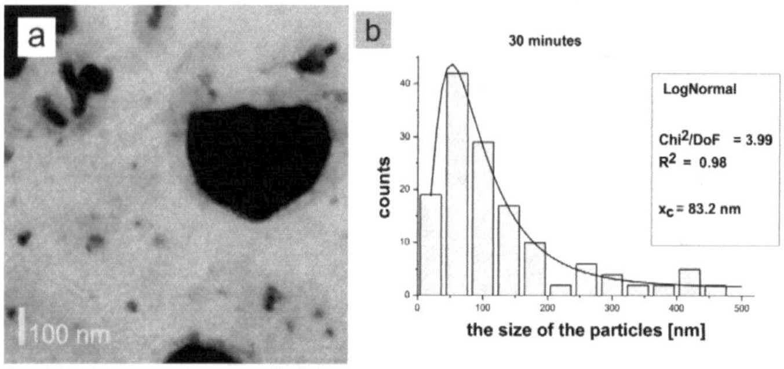

Figure 8 a) TEM image of S30 film and b) histogram of Pd nanograins size distribution.

3.3 Cathodoluminescence studies

In figure 9a and 9b, SEM and CL images for S5 film are presented. On CL image 3 objects which emit light are visible. These objects are the palladium nanograins. In figure 9c and 9d CL spectra collected from two objects (top left and right) are showed. In both spectra two wide emission bands are visible. The first emission band with a maximum at 450 nm and the second one with a maximum at 750 nm are observed. It seems that band at 750 nm of the left object is weaker than one of the right object.

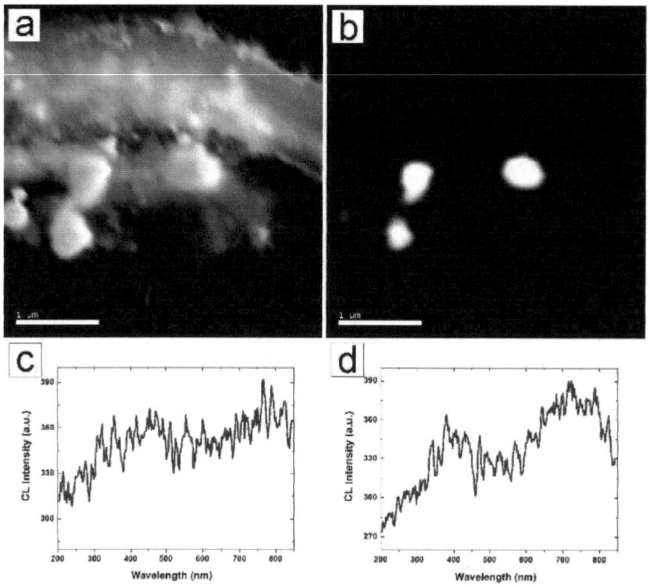

Figure 9 a) SEM and b) CL images of S5 film, c) and d) CL spectrum collected from a top left object and a right object.

In figure 10 SEM and CL images of S10 film are showed. On CL image 2 objects with the luminescence are observed.

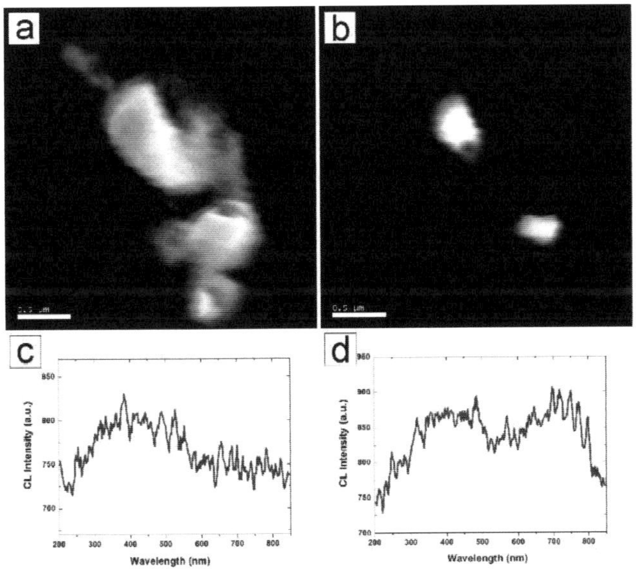

Figure 10 a) SEM and b) CL images of S10 film, c) and d) CL spectrum collected from a top left object and a bottom right object.

In both cases emission band at 450 nm is visible, but a bottom right object also shows a strong band at 750 nm.

In figure 11 SEM and CL images of S30 film are presented. SEM image shows a number of objects with different shapes and sizes. These objects vary the intensity of light on CL image. CL spectra were collected from two objects with the highest intensity. The emission spectrum collected from a top left object is very “indistinct” but spectrum collected from a bottom right object shows strong emission band at 450 nm.

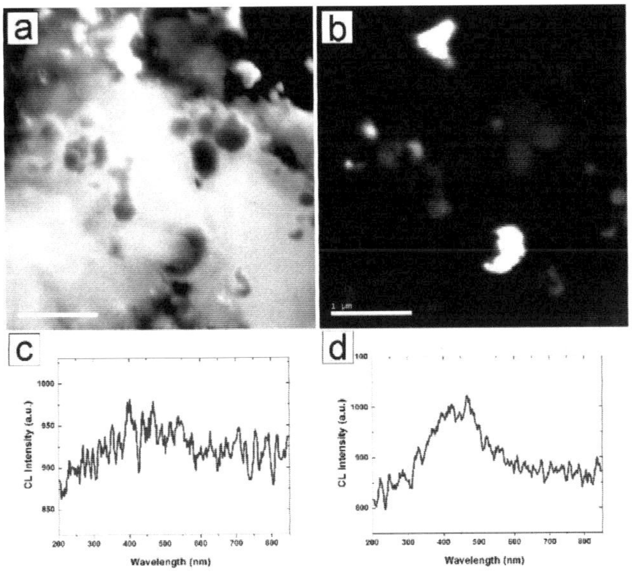

Figure 11 a) SEM and b) CL images of S30 film, c) and d) CL spectrum collected from a top left object and a bottom right object

It results from our previous studies described in paper [16] that Pd nanograins encapsulated with graphite shells are optical active. The shapes and sizes of Pd nanograins may affect the CL spectra and their maxima of emission bands. We suggest that graphite shells may be responsible for the luminescence of Pd nanograins. Further research is required to determine which Pd nanograins have a maximum at 450 nm and which have a maximum at 750 nm. Recently, cathodoluminescence spectroscopy has been used to study the surface plasmon resonance in metallic nanoparticles [23]. According with this paper metallic nanoparticles can absorb and emit visible light due to the surface plasmons.

3.4 Hydrogen sensing of C-Pd films

Hydrogen sensing measurements were performed for S5, S10 and S30 films. The initial resistance of these films measured in air atmosphere is shown in table 2. The films resistance decreased in the function of duration of CVD process. This can be connected with the increase of Pd/C ratio and/or the increase of size of palladium nanograins, what was proved at TEM studies.

Table 2 Resistance of C-Pd films in air

Films	R_0 [Ω]
S5	$2.14 \cdot 10^9$
S10	$1.27 \cdot 10^8$
S30	$7.36 \cdot 10^7$

In the sensor evaluation, the C-Pd films resistance was measured continuously during their interaction with hydrogen. For each sample, measurements for 0.5 vol.%, 1 vol.% and 2 vol.% of hydrogen in nitrogen were performed. The investigations were carried out in continuous gas flow. The atmosphere in the measurement chamber was changed every 10 min – H_2/N_2 and air alternately. The character of changes of samples resistance during hydrogen absorption and desorption was the same for all investigated films. The resistance of all films decreased in hydrogen presence, while the introduction of air caused the increase of films resistance. As a typical result, changes in the resistance of S30 film in 1%H_2/N_2 are shown in figure 12. Additionally, in the right axis of figure 12 the changes of sensitivity versus time are shown. The sensitivity to hydrogen is defined as the percentage change in film resistance. The sensitivity ($\Delta R/R_0$) was calculated using the following equation:

$$\Delta R / R_0 = \frac{R - R_0}{R_0} \cdot 100\% \quad (1)$$

where R - film resistance after exposure to H_2, R_0 - film initial resistance in H_2 absence.

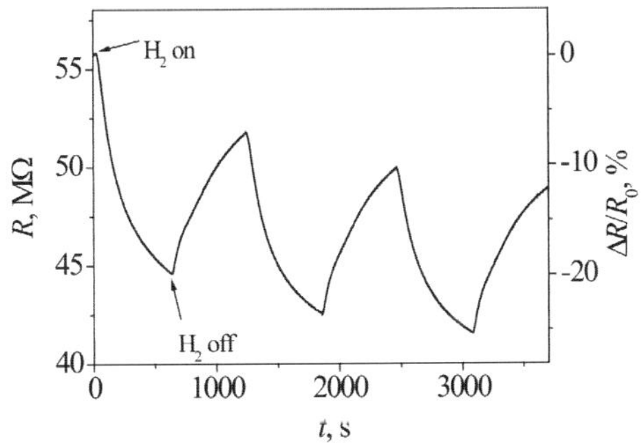

Figure 12 The resistance change of S30 film in mixture of 1%H_2/N_2

The sensitivity of the studied films varied depending of time of CVD process. It also depended on hydrogen concentration. In figure 13 the sensitivity of S5, S10 and S30 films in the function of hydrogen concentration is presented. These values were calculated after 2 min from introduction of hydrogen. As one can see, the highest sensitivity characterized the film modified in 30 min, while the lowest one was observed for S5 film for all studied H_2 concentrations. This means that the application of the longest duration of CVD process led to obtaining of the film with the best hydrogen sensing properties.

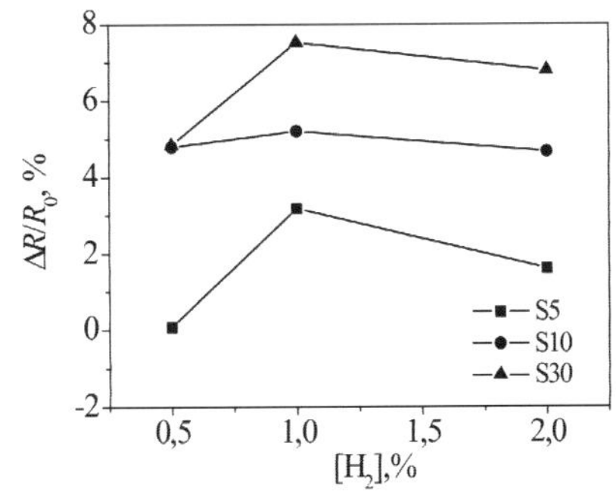

Figure 13 The sensitivity of S5, S10 and S30 films as a function of H_2 concentrations

4. Conclusion

In this work we present the carbonaceous – palladium films after modification by CVD method using various time. We studied the influence of duration of CVD process on morphology, topography and structure of the carbonaceous matrix and shapes and sizes of palladium nanograins. We observe that with increasing duration of CVD modification, the thickness of films decreases but size of palladium nanograins increases. Resistance of film with the smallest thickness is the lowest. This result can be caused by the presence of bigger Pd nanograins with a diameter above 300 nm, which may touch each other in some areas on film's surface. Investigations of hydrogen sensing properties show that film modified in 30 minutes has the highest sensitivity and can be used as an active layer in hydrogen sensor.

In TEM and SEM studies palladium nanograins encapsulated in graphite shells were found. These Pd nanograins showed emission bands with maxima at 450 and 750 nm. We suggest that the graphite shells are responsible for the optical activity of Pd nanograins.

Acknowledgements

This project is co-financed by the European Regional Development Fund within the Innovative Economy Operational Programme 2007-2013 No UDA-POIG.01.03.01-14-071/08-10

References

[1] Kishore S, Nelsonb J, A Adairb, J H Eklund P C 2005 *J. Alloys and Compounds* **389** 234-242
[2] Yürüm Y, Taralp A, Veziroglu T N 2009 *Int J. Hydrogen Energy* **34** 3784-3798
[3] Dibandjo P, Zlotea C, Gadiou R 2013 *Int J. Hydrogen Energy* **38** 952-965
[4] Chen P, Wu X, Lin J 1999 *Science* **285** 91-93
[5] Sayago I, Terrado E, Lafuente E, *et al.* 2005 *Synth. Met.* **148** 15-19
[6] Penza M, Antolini F, and Vittori M A 2004 *Sens. Actuators* B **100** 47-59
[7] Christofides C and.Mandelis A 1990 *J.Appl.Phys.* **68** R1- R30
[8] Lee S W, Yabuuchi N, Gallant B M, *et al.* 2010 *Nature Nanotechnology* **5** 531-537
[9] Khantimerov S M, Kukovitsky E F, Sainov N A 2013 *International Journal of Chemical Engineering* **2013** ID 157098, 4 page
[10] Valentini F, Orlanducci S, Terranova M L 2004 *Sens. Actuators B* **100** 117-125
[11] Kowalska E, Czerwosz E, Diduszko R, *et al.* 2013 *Sens. Actuators A* **203** 434-440
[12] Diduszko R, Kozłowski M, Czerwosz E, Kamińska A, Nietubyć R 2013 *Solid State Phenomena* **203-204** 398-401
[13] Kowalska E, Czerwosz E, Kamińska A, *et al.* 2012 *J.Therm Anal.Cal.* **108** 1017-1023
[14] Czerwosz E, Dłużewski P, Kowalska E, *et al.* 2011 *Physica Status Solidi* C **7-8** 2527-2531
[15] Czerwosz E, Diduszko R, Dłużewski P, *et al.* 2008 *Vacuum* **82** 372-376
[16] Sobczak K, Dłużewski P, Witkowski B P, *et al.* 2012 *Solid State Phenomena* **186** 177-181
[17] Sobczak K, Dłużewski P, Klepka M T, *et al.* 2012 *Int J. Hydrogen Energy* **37** 18556-18562
[18] Pavlovsky I, Soundarrajan P, Yaniv Z 2006 *Sens. Transducers J* **73** 793-798
[19] Schlapbach L and Züttel A 2001 *Nature* **414** 353-358
[20] Vivier Ch C, Bastide S, Laurent M, *et al.* 2012 *Electrochimica Acta* **83** 133-139
[21] http://farmacja.cm-uj.krakow.pl/~mkz/MS-skrypt.pdf
[22] Rymarczyk J., Kamińksa A., *et al.* 2013 *Optica Applicata XLIII No 1 123-132*
[23] Garcia de Abajo F J 2010 *Rev. Mod.Phys* **82** 209-275

Nanocomposite carbonaceous-palladium thin films for ammonia sensors

S. Krawczyk[1] , A. Kamińska[1], M. Kozłowski[1], J. Radomska[1], E. Czerwosz[1] , K. Sobczak[2],
[1]Tele and Radio Research Institute, Ratuszowa 11, 03-450 Warsaw, Poland
[2]Institute of Physics, Al. Lotników 32/46 02-668 Warsaw, Poland

E-mail: slawomir.krawczyk@itr.org.pl

Abstract. Nanostructural carbonaceous-palladium (C-Pd) films for ammonia sensing were studied. These C-Pd films were prepared in a two-step process: the first one involved physical vacuum deposition (PVD) and the second one chemical vacuum deposition (CVD). The structure and morphology of the obtained films were characterized with the use of Scanning Electron Microscopy (SEM) and Transmission Electron Microscopy (TEM). Measurements of ammonia sensing showed that resistivity of the film changes with ammonia content in ambient atmosphere.

1. Introduction

Ammonia is one of the most common chemicals. Due to its high nitrogen content, ammonia is mainly used as a fertilizer. Ammonia also finds its industrial applications, e.g. in cleaning, refrigeration engineering, manufacturing of various chemical products, textile, and finishing. Although ammonia is so widely used it is both caustic and hazardous because of high toxicity. Ammonia gas is dangerous and harmful. Its detection makes an important issue.

There are a lot of sensors for ammonia [1], most of them based on zinc oxide as the active material [2-5] or on titanium oxide [6,7]. They have high power consumption since proper operation of sensors requires the use of heating elements. Ammonia gas is very corrosive, often resulting in a drift and short life of the sensors.

In this paper we propose to apply nanocomposite carbonaceous-palladium films for ammonia sensing. Based on our previous studies [8-11] we know that these films have a highly-developed surface and they are porous.

Specially manufactured carbonaceous-palladium films are sensitive to ammonia and do not require heating. They also demonstrate promising sensory properties for the detection of ammonia. The cost of producing a sensor based on carbonaceous-palladium films and materials for the sensors is low.

The ammonia sensing mechanism could depend on a change in the concentration of charge carriers in the C-Pd films [12,13]. Such an explanation was proposed for palladium/polypyrrole nanocomposite films [14]. Chemisorption of ammonia molecules on the surface of palladium nanograins is accompanied by a charge transfer from NH_3 to Pd, leading to the reduction in the number of holes in the studied films. This causes a change in the film electrical properties. However, a different reaction could take place in the case of C-Pd films. Brown [15] and Allen, Goddard & Schaefer [16] suggested and calculated that the ion-molecule reaction between C+ and NH_3 might produce a significant amount of H_2NC+ ion in addition to the linear HCNH ion as well as H atoms; namely,

$C^+ + NH_3 \rightarrow H_2NC^+ + H$

the metastable H_2NC+ ion can then undergo a dissociative recombination reaction with electrons which, unless there is a considerable rearrangement of the atoms, produces only HNC as well as hydrogen atoms:

$H_2NC^+ + e^- \rightarrow HCN + H$

These hydrogen atoms could penetrate into palladium crystals. The electrical properties of the entire C-Pd film could also change due to both effects.

In this work we present the results of studies on the interaction between ammonia and C-Pd films. We found that the resistance of C-Pd films is increasing in the presence of ammonia gas. We also found that the structure, topography and morphology of the film strongly depends on the temperature of the CVD process.

2. Experimental

The C-Pd films were obtained in a two-step PVD/CVD process that is described in papers [17, 18]. We used fullerene C_{60} (Sigma-Aldrich, 99.9%) and palladium acetate (Sigma-Aldrich, 99.98%) as the film precursors in the PVD process. The same technological parameters were used in all PVD process to prepare samples. The films were deposited on alundum (Al_2O_3) substrate.

The C-Pd PVD films were modified in the second step, in which the pyrolysis of xylene (C_8H_{10}), as the source of carbon, proceeded. The pyrolysis temperature 550°C for sample P1, 650°C for sample P2 and 750 °C for sample P3. The xylene flow and duration of the CVD process were the same for all applied temperatures.

Scanning Electron Microscopy (SEM) and TEM were used to characterize the structure and morphology of the obtained films. SEM examination was carried out with the JEOL JSM-7600F field emission scanning electron microscope operating at 2keV and 5keV incident energy. The structure of the films annealed at 550°C, 650°C and 750°C was investigated with Scanning Electron Microscopy (SEM) in a dual imaging mode: SE (Secondary Electron) imaging and LABE (Low Angle Backscattering Electron) imaging.

TEM investigations were carried out with Titan 80-300 Cubed transmission electron microscope (TEM) operating at 300 keV incident electron beam energy.

A special measurement chamber was used to measure the sensitivity of films for ammonia. The measurement chamber had an electrical feedthrough, gas flow inlet and outlet. The gas flow inlet was driven by a mass flow controller. The flow of NH_3/N_2 mixture (100 ppm NH_3) was set at the level of 50 ml/min. Measurements were carried out at a room temperature and under atmospheric pressure. The desorption of ammonia was performed in the airflow. Electrical resistance of C-Pd films was measured during their interaction with ammonia. Contacts for resistance measurements were obtained by applying conductive silver paint (SPI Supplies) to the films.

3. Results and discussion

3.1. SEM characterization

The structure of films that were obtained in the PVD process varies depending on the sample position in relation to the sources of palladium acetate and fullerene. SEM images show that the films are composed of grains with different shapes and diameters, arranged layer by layer on the substrate. The structure of the film suggests that grains grow in a form of independent islands. These islands are in a form of large 200 nm grains (Figures 1a, 2a, 3a). The film topography changes due to the surface filling with the islands.

The structure of films modified with CVD changes with the annealing temperature. Pd nanograins in the C-Pd layer agglomerate due to annealing. For C-Pd films obtained at 550°C and 650°C the effect of Pd nanograins agglomeration can be observed (Figures 1b and 1c, Figures 2b and 2c). The newly formed grains are shown as bright objects in a SEM image in the LABE mode (Figures 1c and 2c). These Pd grains have a diameter of a few hundreds nm. On the surface of the film modified at 550°C and 650°C we observe many randomly distributed islands of a porous carbonaceous structure (Figures 1d and 2d) and small Pd grains (Figures 1c and 2c). Such superficial Pd grains can be seen as bright objects in Figure 1c and 2c. For C-Pd film modified by CVD at 550°C we observe that the carbonaceous islands are coated with Pd nanograins (Figure 1d).

Smaller Pd grains are observed for a film annealed at 750°C (Figure 3b, c). These Pd grains are situated on the edges of the substrate huge grains (> a few micrometers). On the flat fragments of the film surface, Pd nanoparticles with a diameter of ca. 1 micron are clearly visible at high magnification (Figure 4a, 4b). Besides these large grains, a lot of very fine Pd nanograins are found on the film surface. These nanograins are densely and uniformly distributed on the surface.

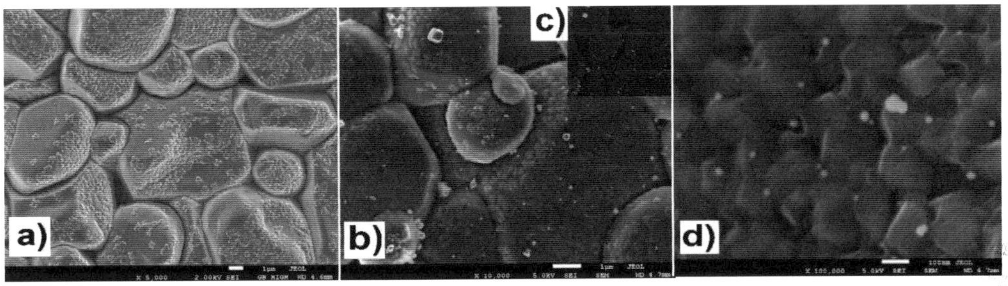

Fig. 1. SEM images of sample P1 surface a) after PVD process, b) after CVD process (temp. 550°C) SE imaging, c) after CVD process (temp. 550°C) LABE imaging, d) SE imaging of a fragment of a carbonaceous island coated with Pd nanograins

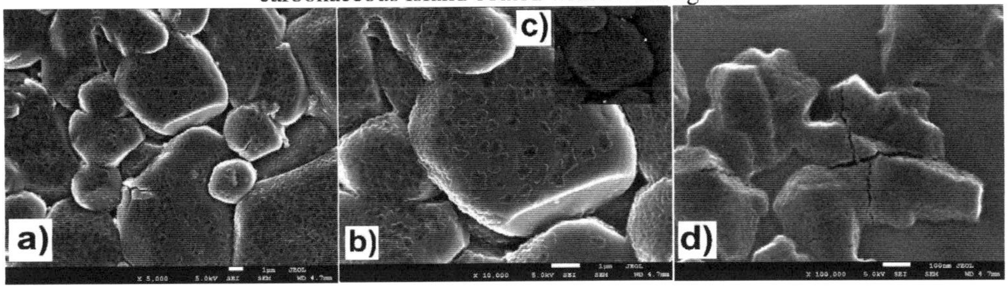

Fig. 2. SEM images of sample P2 surface a) after PVD process, b) after CVD process (temp. 650°C) SE imaging, c) after CVD process (temp. 650°C) LABE imaging, d) SE imaging at 100,000 magnification, showing a porous island

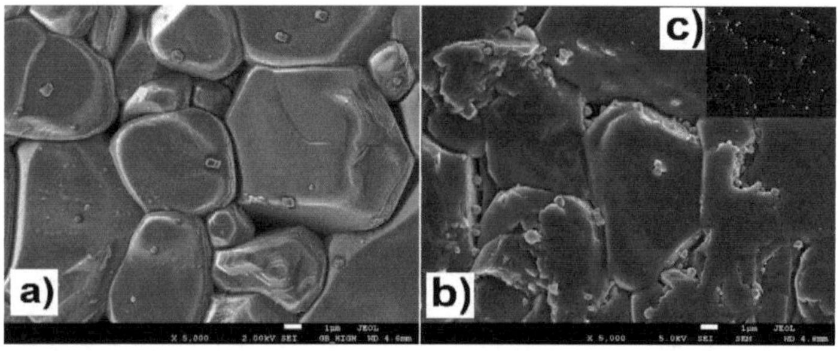

Fig. 3. Images of SEM of sample P3 surface a) after PVD process, b) after CVD process (temp. 750°C) SE imaging, c) after CVD process (temp. 750°C) LABE imaging

Fig. 4. . Images of SEM of sample P3 surface a) after CVD process (temp. 750°C) SE imaging, b) after CVD process (temp. 750°C) LABE imaging

TEM images of the annealed films are presented in Figures 5a, b and c. For the C-Pd film modified at 550°C only small Pd nanograins (5-10nnm) with *fcc* structure type are observed. The structure of the Pd grains was determined based on the analysis of the electron diffraction pattern. A higher temperature of CVD process causes a formation of larger Pd grains, with sizes significantly exceeding 200 nm for 650°C and with sizes up to 150 nm for films obtained at 750°C. It should be noticed that Pd nanograins, observed in films obtained at 750°C, are surrounded with a carbon shell. It shows in Fig. 5c, where the dark objects are Pd nanograins and a lighter shell can be seen around each Pd nanograin.

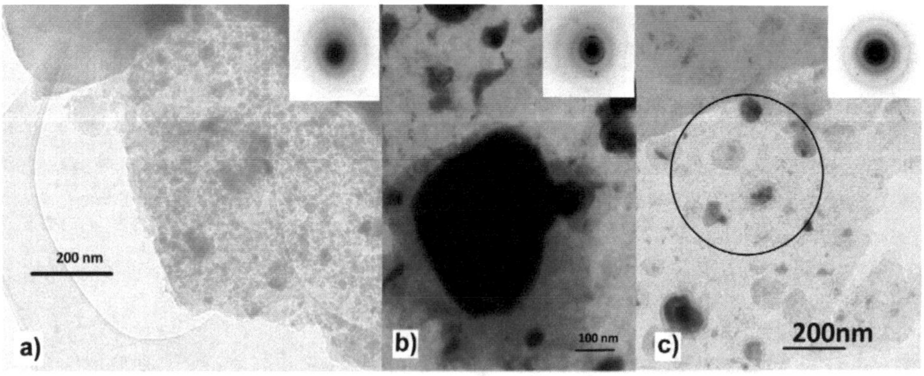

Fig. 5. Images of TEM of C-Pd films obtained at: a) 550°C, b) 650°C and c) 750°C. The circle in Figure 5c marks the area with Pd nanograins surrounded with carbon shells. The insert presents the electron diffraction pattern for relevant films.

3.2. *Sensing Measurements*

The sensing measurements for annealed C-Pd films are based on changes in the film resistivity in ambient atmosphere containing ammonia. These changes are analysed according to the following formula:

$$\Delta R / R_0 = \frac{R - R_0}{R_0} \cdot 100\% \qquad (1)$$

where R stands for the sample resistance in the test gas and R_0 stands for the sample resistance in the air.

The resistivity of the films increased after introducing ammonia to the measurement chamber, while the initial resistivity of these film was 2-5•10^3 Ω • m. Such resistivity value could be attributed to the presence of many porous carbon islands. Kennedy et al. [19] studied resistivity of such materials and they found that this resistivity was $10^2 - 10^3$ Ω •m.

Introduction of air into the measuring chamber causes a decrease in resistivity of samples but did not restore the initial resistivity value. Figures 6a, b and c present the results obtained for the three investigated samples P1, P2 and P3 respectively.

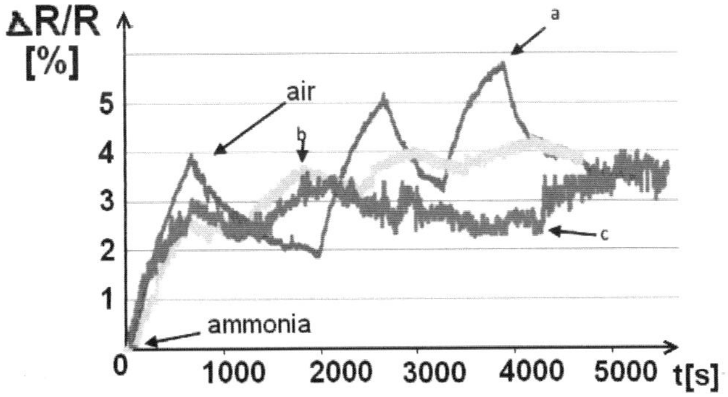

Fig. 6. Sensitivity of C-Pd films *versus* time of exposure to 100 ppm NH_3: a - sample P1, b - sample P2, c - sample P3.

It should be noticed that P1 demonstrated the best response for ammonia, while P2 revealed the worst one. Taking into account the mechanism of electrical change of the C-Pd film during an interaction with ammonia suggested in the introduction, we can suppose that the presence of a porous carbonaceous structure, rich in C^+ ions is the most important factor in the process of ammonia decomposition and hydrogen incorporation into palladium grains. This leads to a formation of PdH_x whose resistivity is higher than the resistivity of palladium. Consequently, the film resistivity during ammonia absorption increases.

4. Conclusions

Nanocomposite carbonaceous-palladium (C-Pd) films, prepared at the Tele and Radio Research Institute, make promising materials for ammonia sensor applications. The films were characterized with SEM (after PVD and PVD/CVD process). Ammonia sensing properties were also measured for all these films. Our SEM investigation of the initial film obtained from the PVD process reflected the substrate surface shape, which is composed of angular grains of few hundreds nm in size. Microscopic studies of the PVD/CVD films revealed that the film topography and morphology strongly depends on the temperature of CVD process. The best sensing properties were obtained for films where Pd nanograins formed conducting paths on the edge of large grains of a substrate covered with C-Pd film and many porous carbon islands were located on the film surface. The worst sensing properties were observed for a film without such conducting paths. In the film annealed at 750°C, Pd nanograins were surrounded with carbon shells that impeded hydrogen desorption.

Acknowledgements - This project is co-financed by the European Regional Development Fund within the Innovative Economy Operational Programme 2007-2013 No UDA-POIG.01.03.01-14-071/08-09.

References

[1] Timmer B, Olthuis W and Van den Berg A 2005 Sens. Act. B **107** 666–677
[3] Hidehito N, Tadatsugu M and Shinzo T 1986 *Appl. Phys.* **60** 482-484
[3] Patil A, Sonawane L and Patil D Res. 2011 *J.Mod.Phys.* **2** 1215-1221
[4] Moseley P 1992 *Sensors and Actuators B*, **6** 149-156
[5] Pizzini S, Butta N, Narducci D and Palladino M 1989 *J.Electrochemical Soc.* **136** 1945-1948
[6] Karunagaran B, Uthirakumar P, Chung S, Velumani S and Suh E 2007 *Mater. Charac.* **58** 680–684
[7] Pawar S, Chougule M, Patil S, Raut B, Dalvi D , Patil P, Sen S, Joshi P and Patil V 2011 *Sensor J. of Sens. Tech.* **1** 9-16
[7] Mehtaa R, Andersonb D and Hagerc J 2003 *Carbon* **41** 2174-2179
[8] Chen C, Kennel E, Stiller A, Stansberry P and Zondlo J 2006 *Carbon* **44** 1535-1543
[9] Beechem T, Lafdi K 2006 *Carbon* **44** 1548-1559
[10] Rode A, Hyde S, Gamaly E, Elliman R, McKenzie D and Bulcock S 1999 *Appl. Phys.* **A** 755-758
[11] Zhang D and Zhang W 2009 J. Sens. 1-16.
[12] Saroch M, Srivastava S, Fink D and Chandra A 2008 *Sensors* **8** 6355-6370
[13] Schönauer-Kamin D, Fleischer M and Moos R 2011*Sensors* **13** 4760-4780
[14] Hong L, Li Y and Yang M 2010 *Sensors and Actuators B* **145** 25–31
[15] Brown R. D. 1977 *Nat* **39** 270
[16] Allen T, Goddard J and Schfefer III H 1980 *J. Chem. Phys.* **73** 3255
[17] Kozłowski M, Czerwosz E, Dłużewski P, Kowalska E, Radomska J and Wronka H 2009 *JAMME* **37** 304.
[18] Choy K 2003 *Progr. Mat. Sci.* **48** 57–170
[19] John Kennedy L, Vijaya J and Sekaran G 2005 *Mat. Chem. and Phys.* **91** 471–476

Journal of Physics: Conference Series **564** (2014) 012005 doi:10.1088/1742-6596/564/1/012005

Temperature modulation in dilatometric investigation of PVD substrate-coating systems

P Myśliński[1], P Kamasa[2], A Gilewicz[1]

[1]Koszalin University of Technology, Institute for Technology and Education

75-435 Koszalin, ul. Śniadeckich 2, Poland

e-mail: piotr.myslinski@tu.koszalin.pl

[2]Wigner Research Centre for Physics, Research Institute for Solid State Physics and Optics

Hungarian Academy of Sciences

1125 Budapest, POB 49, Hungary

Abstract

Temperature modulation technique and phase sensitive analysis is applied for investigation the thermal stresses induced at higher temperature in the space between coating and substrate. The method applying only phase shift parameter eliminates influence of the intensity variation of measured signal. Comparison with conventional application of temperature modulation is presented. Method gives comprehensive information about the state of coating which is usually immersed using amplitude analysis.

1. Introduction

Intrinsic material properties such as hardness, strength, ductility, work hardening etc. are very important factors for wear-resistance of the working surface. But also other factors associated with surface properties like surface finish, lubrication, load, corrosion temperature etc. are equally important. Enhance of a wear resistance can be obtained by coating the working surface by ceramic thin films. Ceramics like TiN, TiAlN TiC, TiCN or CrN are characterized by high wear-resistance. Among other methods of deposition, the plasma assisted physical vapor deposition (PAPVD) is often used. Beside the surface properties, the resistance on a mechanical and thermal load decides about usability of such modification. This depends mainly on the adhesion between coating and a substrate as an intermolecular interaction. The adhesion is seriously reduced by contribution from extraneous sources [1]. These include internal stresses, which are almost always present in the layers deposited on a substrate. One sort of stress originates from the different lattice parameters of layer and substrate (coherency stresses). During process of coating intrinsic stress may be generated when the layer grows [2]. Thermal stress is a result of different thermal expansion coefficient of layer and substrate when temperature is changing after the process is completed [3]. At higher temperatures the chemical and structural processes are activated and adhesive forces can be reduced and thermal stability decreased. The degradation (delamination) of the coating occurs when the adhesive forces are reduced by forces from all kind of stresses, among which the thermal stresses dominate at high temperature.[1]

[1] To whom any correspondence should be addressed.

Journal of Physics: Conference Series **564** (2014) 012005 doi:10.1088/1742-6596/564/1/012005

In the field of thin wear resistance coatings the temperature stability of the adhesion is the basic pa[2]rameter in the industry application. A method for the diagnosis of the state of a coatings obtained by physical vapour deposition on tools for metal and wood processing has been developed which is based on recording thermomechanical effects at changing temperatures. These thermomechanical effects reflect the mechanical stresses between substrate and coated thin layer, which depend on physical properties of the materials, mainly on thermal expansion and compressibility. These interactions can be detected by dilatometry as changes in effective thermal expansion. The effective thermal expansion of the substrate-coating system with ideal adhesion can be estimated theoretically and than verified experimentally. Any deviation from the ideal state of the deformation can be interpreted as loss of adhesion. It was recognized that a linear temperature program is not sensitive enough when the sample is massive as in case of a cylindrical rod as described in *Experimental* section of this paper.

A major advance in dilatometric (DIL) experiment is the application of a temperature modulation (TM), leading to temperature modulated dilatometry, TMDIL [5, 6]. In general, temperature modulation (TM) enables recording of at least two parameters of dilatometric response. Using the sinusoidal temperature modulation enables first of all recording the response as intensity of the mechanical dilatation A_{DIL}, independent of the phase changes of the signal. This gives increased sensitivity in detection changes of the coefficient of thermal expansion (CTE) [7]. Another experimental aspect of TM is the pure analysis of phase changes which enables to obtain additional useful information about delamination processes of wear resistance coating. A new method developed by authors brought promising experimental results and is presented here as a preliminary experimental achievement.

2. Experimental

2.1. Sample. Titanium nitride (TiN) films were deposited on pure iron (ARMCO) substrates by the PVD method using cathodic arc evaporation (CAE). The Titanium cathode was 100 mm in diameter. The arc current in the process of deposition was 80 A. The gas atmosphere in the working chamber was pure nitrogen at a pressure of 0.5 Pa and the substrate bias voltage was –70 V. The substrate temperature during film deposition was controlled to a value of about 400°C. The distance between the substrate and the Ti cathode surface was 18 cm. The substrate was in the shape of a cylindrical rod, 30 mm in length and 3 mm in diameter. The average film thickness was 2.4 μm and the deposited film was homogenous in thickness within ±0,3 μm. In order to get a homogeneous film the rod sample was rotated during deposition.

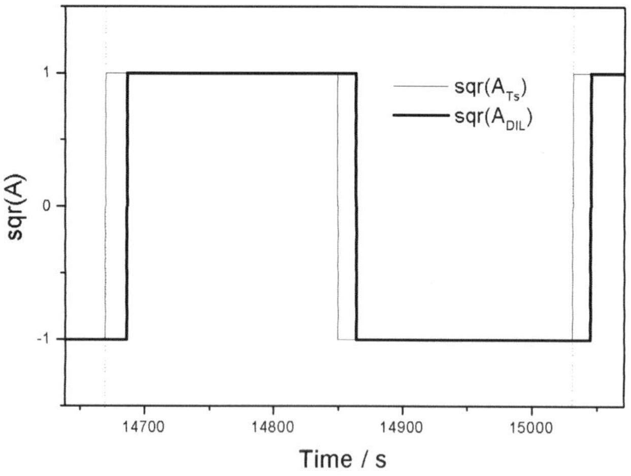

Figure 1. Square waves obtained from sample temperature and dilatation signals, period 360 s.

Journal of Physics: Conference Series **564** (2014) 012005 doi:10.1088/1742-6596/564/1/012005

2.2. Phase detection. In this work a new experimental aspect is realized, namely applying temperature modulation together with analysis of the phase shift $\phi_{DIL/Ts}$ of the A_{DIL} signal with respect to the phase of the sample temperature signal, A_{Ts}. The method will be described in details in forthcoming paper [8]. Principles have been adopted from radiofrequency techniques using FM modulation and detection system. Briefly, analog sinusoidal signals are transformed electronically into the square waves being in phase with sinusoids: sqr(A_{Ts}), sqr(A_{Tp}), sqr(A_{DIL}) corresponding to sample temperature, program temperature and dilatation, respectively. Amplitude is limited to the unit and zero-transitions determine the phase shift $\phi_{DIL/Ts}$ of the A_{DIL} signal versus A_{Ts} (figure.1).

It is known from earlier investigations [9] that the presence of a thin layer mechanically interacts with the substrate. For the model sample in shape of rood the mechanical interactions lead to physical deformation with temperature (stretching or shrinking in length) measurable using dilatometer and easily to be analyzed numerically. Finally the effect of interest is that loosing of the sample coating adhesion appears as nonreversing axial local relocations of the coated layer versus substrate. With sinusoidal variation of the temperature the relocations of the layer change periodically their direction to the opposite (hysteresis). The effect is observed as an additional phase shift between sample temperature A_{Ts} and dilatation Λ_{DIL}. When thin film has ideal adhesion to the substrate the phase shift appears as a result of heat diffusion to the sample and the resulting gradients. It should be noted that temperature modulation in spite of thermomechanical effects, enables to detect dilatometric effects associated with thermodynamical changes of the sample, i.e. reversible and irreversible transitions [10].

2.3. Uncoated Fe sample. Phase analysis in comparison with DIL amplitude variation for uncoated Fe sample in shape of rod is illustrated in figur 2. The result was obtained for sinusoidal modulated temperature with 5°C of amplitude and a period of 50s. In this experiment applying temperature modulation, three values were simultaneously recoded: sample temperature T_s, dilatation DIL and independently the magnetic susceptibility presented as TMAG signal.

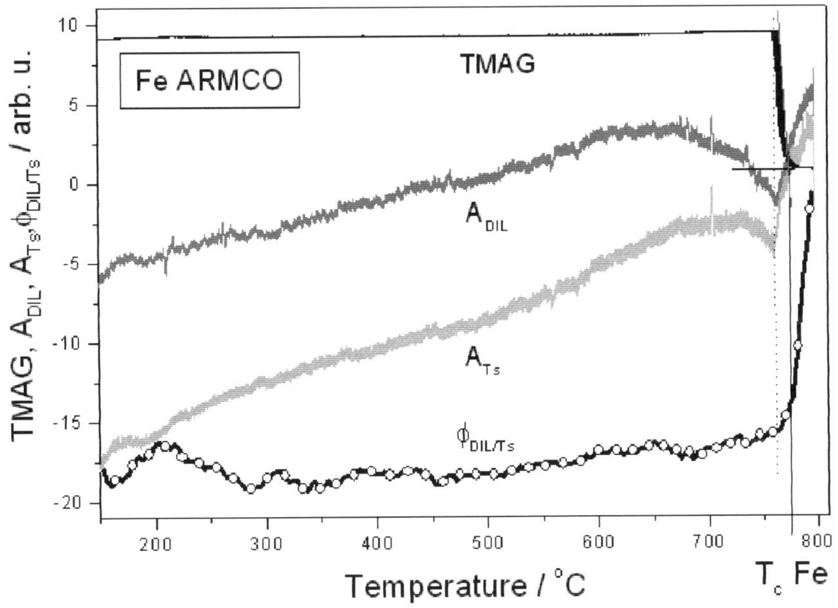

Figure 2. Temperature modulated method applied to the Fe rod sample: The phase shift $\Phi_{DIL/Ts}$ analysis by phase sensitive procedure in comparison with conventional amplitude analysis of A_{DIL}, A_{Ts} as well as TMAG. The reader should keep in mind that the Curie point is at 770°C.

From traces of DIL and T_s it can be concluded that changes in the DIL amplitude A_{DIL} are mainly a result of temperature. The trace of temperature amplitude variation A_{Ts} is nearly the same (besides of a constant factor) as dilatation amplitude variation A_{DIL}. In contrast, the phase variation $\phi_{DIL/Ts}$ is significantly associated with changes of magnetic properties of the sample seen above the Curie temperature of iron.

3. Results

The specific behavior of a TiN layer on an pure iron (ARMCO) substrate was investigated using a model sample in shape of cylindrical rod. Results are presented in figure 3 separately for amplitude (left column) and phase analysis (right column).

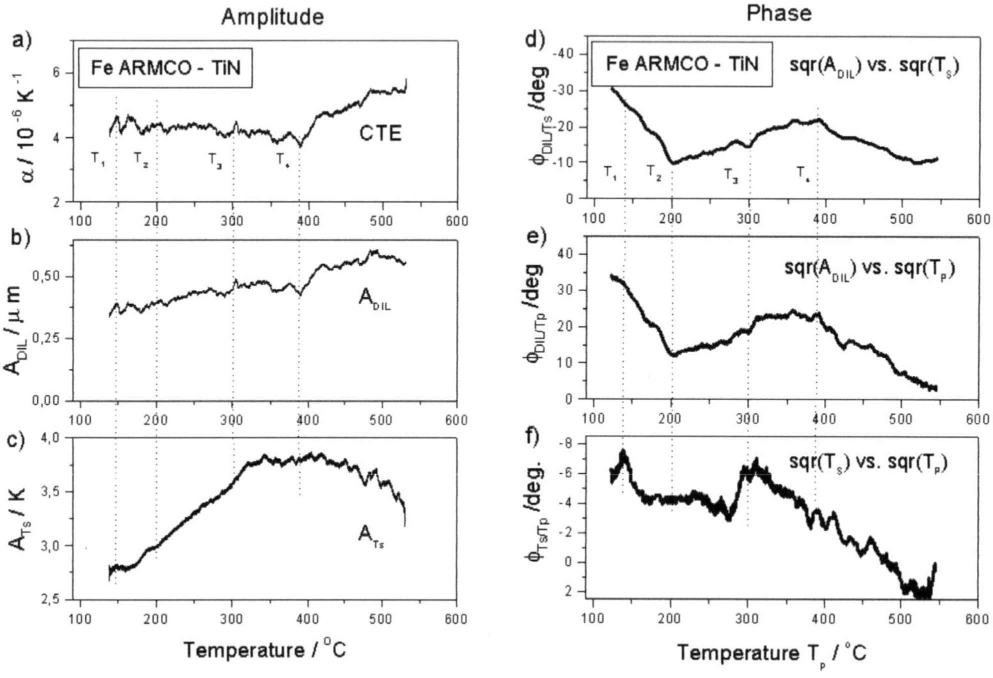

Figure 3. Result of phase detection (right) in comparison with traditional amplitude analysis (left). Three cases of phase analysis are available: sqr(A_{DIL}) phase changes in relation to the sample temperature sqr(Ts) as reference (d), sqr(A_{DIL}) phase changes in relation to the program temperature sqr(Tp) as reference (e), sqr(Ts) phase changes in relation to the program temperature sqr(Tp) as reference (f).

The amplitude analysis enables to obtain coefficient of thermal expansion (CTE) from the amplitude of dilatation, A_{DIL}, and the sample temperature, A_{Ts}, respectively. This value has an analogue meaning as direct current measurements with features for identification of the reversible and irreversible processes.

In case of phase detection there are three possibilities of analysis: (a) phase changes of sqr(A_{DIL}) using as a reference the sample temperature sqr(Ts) or (b) programmed temperature sqr(Tp), and (c) phase changes of sqr(Ts) using programmed temperature sqr(Tp) as a reference.

In case of periodic changes of T_s, dilatation A_{DIL} changes are connected only with the sample temperature T_s resulting in variation of the phase-shift without intensity variation (figure 3d). The influence of the amplitudes of both signals is minimized since the effect of coating can be distinguished from physical and chemical reactions. For example, in Phase column of figure 3f at temperature T_3 a strong effect associated with substrate being at this temperature during deposition can be seen. In the left part of figure 3 the temperature activation of chemical degradation of the thin film of TiN coating (oxidation) is seen on phase change of A_{DIL} signal at T_4 (see figure 3 d, e).

Conclusions

The applied pure phase analysis to the temperature modulation technique in the field of wear resistance coating technology is promising giving additional comprehensive information. Concluding it can be stated that this method allows not only to discriminate reversible and irreversible processes but more, the method enables to distinguish between irreversible processes ~~as~~ like early coating delaminating and structural phase transitions. The technique used together with all previous application is a complementary one and extends the methods known as Dynamic Load Thermomechanical Analysis (DL TMA).

Reference

[1] Mittal K L, *Electrocomponent Science and Technology*, **3** (1976) 21.
[2] Cammarata R C, Bilello J C, Greer A L, Sieradzki K, Yalisove S M, *MRS Bull.* **4** (2) (1999) 34.
[3] Wendler B, Młotkowski A, Proc. 12th International Summer School Modern Plasma Surface Technology, Technical University of Koszalin, Poland, (2000) 217.
[4] Hultman L, *Vacuum*, **57** (2000) 1-30.
[5] Reading M, Elliot D, Hill V I, *J. Therm. Anal. Cal.* **40** (1993) 949.
[6] Price D M, *Thermochim. Acta* **315** (1998) 11.
[7] Kamasa P, Myslinski P, Pyda M: Thermal expansion coefficient determination by temperature-modulated dilatometry. In: *Proc. 31st Annual NATAS (North American Thermal Analysis Society) Conference* (Albuquerque, NM, USA, 2003). Ed. M.J. Rich (Michigan State University, MI, USA, 2003), on CD-ROM; published also in NATAS Notes, Fall 2003, Vol.35 No.3 pp. 17-21.
[8] Myslinski P, Kamasa P, Phase analysis in temperature modulated thermal analysis, *in preparation.*
[9] Myslinski P, Investigation of the thermal stability of the hard coatings by Modulated Temperature Dilatometry. *Vacuum* (2009) 757÷760.
[10] Wunderlich B, Thermal Analysis of Polymeric Materials, Springer-Verlag, Berlin 2005, p.359.

Journal of Physics: Conference Series **564** (2014) 012006 doi:10.1088/1742-6596/564/1/012006

A cryogenic distillation column for the XENON1T experiment

S Rosendahl[1], E Brown[1], I Cristescu[2], A Fieguth[1], C Huhmann[1], M Murra[1] and C Weinheimer[1]

[1] Institut für Kernphysik, Wilhelm-Klemm Strasse 09, 48145 Münster, Germany

[2] Karlsruher Institut für Technologie, Tritium Laboratory, Hermann Von Helmholtz-Platz 1, 76344 Eggenstein-Leopoldshafen, Germany

E-mail: rosendahl@wwu.de

Abstract. The XENON collaboration aims for the direct detection of cold dark matter in form of weakly interacting massive particles (WIMPs). A dual phase time projection chamber filled with liquid xenon is used to detect the WIMP-nucleon interaction. For the next generation experiment XENON1T with an active target mass of 1 ton of xenon, a new distillation column to remove krypton out of xenon to a concentration of $< 5 \times 10^{-13}$ (0.5 ppt) natural krypton in xenon is designed and tested at the Institut für Kernphysik, Universität Münster. The experimental setup together with two diagnostic tools is presented, as well as one stability test of a 11 hour distillation run at the designed flowrate of 3 kg per hour.

Keywords: Xenon; Krypton; Cryogenic distillation; Ultra-pure noble gases; Dark matter

1. Introduction

A various number of astronomical observations implies, that the matter content of the universe is dominated by non-luminous, non-baryonic dark matter, while the baryonic matter only makes less than 20% of the matter [1]. One candidate for the dark matter, coming from particle physics, is the WIMP, a weakly interacting massive particle. A well motivated WIMP particle is described by super-symmetric enhancements of the standard model of particle physics, which is the lightest supersymetric particle (LSP or neutralino). In different models the neutralino is supposed to be stable and allows the interaction with ordinary matter on the weak scale [2]. In the recent years different experiments tried to detect the interaction between cosmic WIMPs and baryonic matter, looking for energy deposition in earth bound detectors.

The XENON collaboration, searching for detecting WIMP-nucleon interactions, uses a dual-phase time projection chamber filled with xenon, while the xenon itself is the target-mass for the interaction [3]. The detection principle is shown in figure 1. The interaction of an incoming particle with the xenon is producing prompt scintillation light, called S1, as well as electrons in the liquid xenon. The charges are drifted in an electric field to the gas phase. In a second, much stronger field the electrons are extracted into the gas phase and accelerated, producing a second light signal, called S2. The ratio of S1 and S2 allows to discriminate between electronic recoils coming from interactions with electrons or gammas and nuclear recoils that are produced by neutron scattering or the WIMP-nucleon interaction.

Currently the XENON collaboration is running the XENON100 experiment in the Gran Sasso Underground Laboratory, using 62 kg of xenon as detector material, while in parallel the next

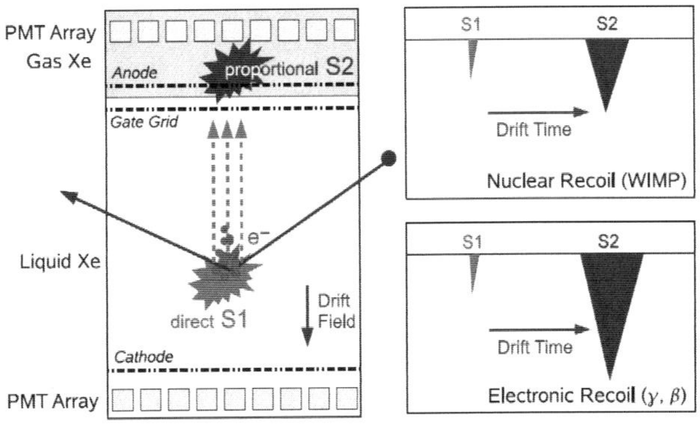

Figure 1. Working principle of the XENON100 TPC. This diagram shows how the signals are generated in the XENON100 dual phase TPC. An incoming particle produces charge and light in liquid xenon (S1), while the charges are guided in an electric field to the gas phase, where they are extracted to produce a second light pulse (S2). The photons are detected using photomultiplier tubes. The ratio of the S2 signal to the S1 signal allows to distinguish between nuclear recoil (WIMP) events and electronic recoil events [3].

generation, called XENON1T with an active detector mass of 1 ton is being build up in the same underground lab. XENON100 was the leading experiment for direct dark matter search, setting the most stringent limit in 2012 for spin-independent WIMP-nucleon cross section of 2×10^{-45} cm^2 at 55 GeV/c^2 and 90% confidence level [4]. This result has recently been confirmed by another liquid xenon experiment, LUX, which has used an increased sensitivity to set a limit of 7.6×10^{-46} cm^2 at 33 GeV/c^2 with 90% confidence level [5].

The XENON1T experiment is designed for a sensitivity to the WIMP-nucleon cross section of 2×10^{-47} cm^2, which is more than one order of magnitude better sensitivity than current experiments. To achieve this goal, the radioactive background has to be reduced compared to actual experiments. Beside investigation of the proper detector materials, it is also necessary to clean the xenon itself from intrinsic radioactive contaminations. One of them is the ^{85}Kr isotope, which is a β-source with an endpoint energy of 687 keV and a half-life of 10.76 years. For XENON1T it has been calculated, that the concentration of natural krypton in xenon has to be reduced below 0.5 ppt (10^{-12} mol/mol), where its decay does not contribute to the background.

2. Separation of krypton and xenon using cryogenic distillation

To reach the needed concentration of natural krypton in xenon, the technique of cryogenic distillation is used. Due to the different vapor pressures of krypton and xenon at a fixed temperature, e.g. -98°C, a separation can be achieved, while the xenon gets liquefied and the krypton stays in the gas phase.

To achieve a certain separation, a series of distillation stages can be implemented. The number of distillation stages can be calculated using the McCabe-Thiele method [6]. Commercially available xenon of highest purity has a concentration of 10 ppb (10^{-9} mol/mol) to 1 ppm (10^{-6} mol/mol) which implies, that a separation factor of 10^4 to 10^6 is needed to reduce the overall amount of krypton in xenon down to 0.5 ppt.

To reach that high separation factors it has been calculated that ~ 10 theoretical distillation stages are needed. For the technical realization it has been decided to construct a package

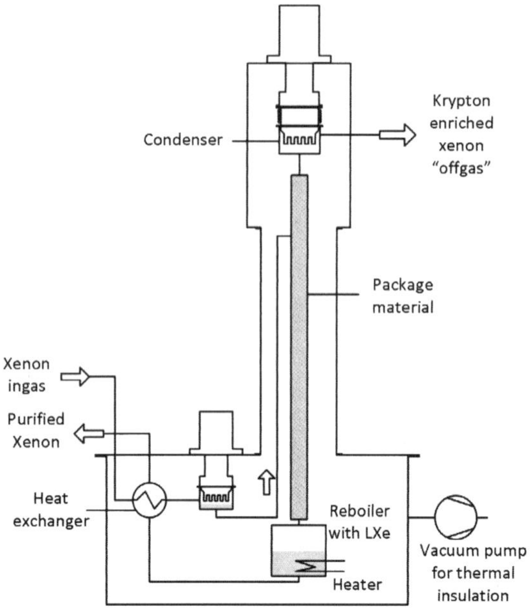

Figure 2. Package type distillation column. Left: Schematics of a package tube distillation column. The key components are the condenser and the reboiler, with the package tube in between, where the xenon is injected and the distillation takes place. Right: Package material, filled into the segments of the tube.

column and to use structured package material (company Sulzer, type EX), as shown in figure 2 (right) to achieve a huge surface, where the distillation process takes place.

The amount of package material, that is needed, can be estimated from the height equivalent value for one theoretical distillation stage (HETP value) multiplied with the number of distillation stages. For this column a package height of $\sim$3 m has been estimated, also including a huge uncertainty in the HETP value, since it is unknown for krypton-xenon mixtures at low concentrations. The schematic view of a package type column is also shown in figure 2 (left). In the bottom, a reservoir is mounted, where the liquid xenon is collected, the so called reboiler, while on the top of the package material a condenser is mounted which supplies a certain cooling power to the system. By heating in the bottom and cooling in the top, a stream of xenon through the package is produced. On the top of the column, the krypton enriched xenon is collected. A small fraction, called offgas, is extracted, while most of the xenon is liquefied again and streaming down to the package (reflux). The purified xenon is extracted from the liquid in the reboiler. The column for XENON1T is designed for a throughput of 3 kg/h with a krypton enriched offgas flow of only 1%. The reflux ratio of 191 is quite high to ensure a good separation efficiency.

3. Technical design of the distillation column

Since there are a lot of open questions, concerning the HETP values and the McCabe-Thiele method for low concentration-separation, it has been decided to split the project in two phases. Phase 1 foresees a distillation plant with only 1 m package material inside, with the advantage of a faster stabilization, which allows to run several performance tests on a shorter time scale, to answer some of the unknown issues.

The incoming xenon is liquefied in a first condensation station and is then injected to the package. The cryo cooler (Typ Leybold CP140T) in top of the column is providing a cooling

Figure 3. Phase-1 Distillation Column. This picture shows the final setup for the 1-m distillation column as it was planed for the first performance tests of the improved distillation design for XENON1T.

power of 200 W at -98°C, to liquefy the upstreaming xenon. The purified xenon is extracted from the liquid in the reboiler and passes a heat exchanger to evaporate and to pre-cool the incoming gas. The krypton enriched offgas is extracted at the top. The Phase-1 setup can be seen in figure 3.

In order to avoid a contamination of the xenon, the system has to be extremely leak-tight and clean. Therefore, all components have been designed to fulfill ultra-high vacuum criteria. All parts have been made of stainless steel and OFHC copper. The pipings are done with electro polished pipes, using VCR connectors, CF flanges or orbital welding technique for leak-tight connections and all parts have been cleaned using ultrasonic cleaning. Before filling any xenon to the system, the setup has been pumped, using an oil-free turbo-molecular pumping station with a scroll pump and especially the package tube with the large surface has been baked out to +100°C to remove impurities. In a second cleaning step, xenon has been circulated through the system and further impurities have been flushed out and removed by the usage of a hot zirconium based getter. The whole system is equipped with various pressure and temperature sensors (PT1000 and calibrated silicon diodes) to monitor the stability of the distillation process.

Journal of Physics: Conference Series **564** (2014) 012006 doi:10.1088/1742-6596/564/1/012006

4. Diagnostics

To determine the efficiency of the distillation process, different approaches are under investigation. In the first method, a system has been designed to measure trace amounts of natural krypton in xenon using a liquid nitrogen coldtrap to enhance the sensitivity of a commercial residual gas analyzer based on a quadrupole mass filter. In the second approach, xenon is doped with a radioactive ^{83m}Kr tracer at the inlet of the column and the activity at the outputs is detected. Both approaches will be briefly presented.

4.1. Detection of krypton in xenon using a quadrupole mass filter

In order to determine the separation efficiency of the distillation plant one has to measure the concentration of krypton at the inlet of the column and at the purified outlet. Since this column works in a regime of very low concentrations of krypton in xenon (sub-ppb range), commercial residual gas analyzers are not able to detect them in the vicinity of the xenon atoms. Therefore, the sensitivity has to be increased. In addition to the distillation plant, a system has been designed, which allows to measure krypton in xenon to a concentration of 40 ppt following an idea of reference [7]. The principle is briefly explained in the following:

To enhance the sensitivity we use again the effect of vapor pressure for the different gases at cryogenic temperatures. The major amount of the xenon sample is frozen out in a liquid nitrogen coldtrap, while the traces of krypton stay in the gas phase and pass the coldtrap nearly unattached. This gas has an increased krypton to xenon ratio and is guided through a differential pumping section into the analyzing chamber, where it can be measured by a commercial rga.

The details of the system are presented in [8].

4.2. ^{83m}Kr-Tracer method to investigate the distillation performance

Beside the rga method, a second technique is developed, which allows a continuous monitoring of the distillation process. Since the rga method only allows to take dedicated samples and analyze them on a timescale of ~1 hour, it is not feasible to monitor the dynamic of the system continuously over a long time period.

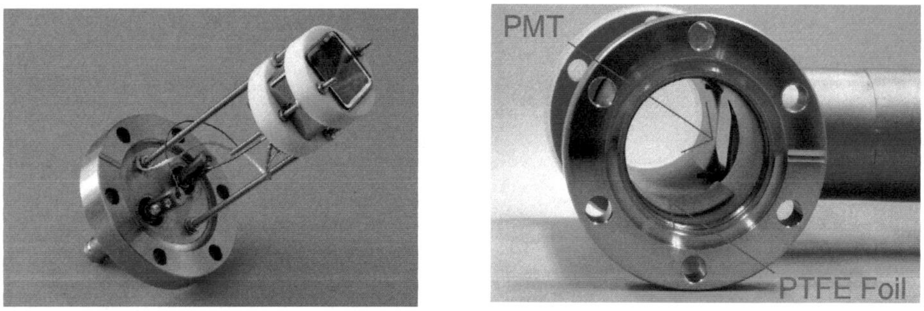

Figure 4. ^{83m}Kr decay detector. Left: Photomultiplier tube mounted with a Teflon holder to a CF-40 feed-through flange to provide high voltage and signal read-out. Right: PMT finally positioned inside a CF-40 T-piece, where the xenon is flushed through in front of the detector.

The idea is to inject trace amounts of the radioactive krypton isomere ^{83m}Kr to the xenon at the inlet of the column and to monitor the decays at both outlets of the column. Since ^{83m}Kr decays with a half-live of 1.8 h to ^{83}Kr, which is stable, there is no danger of spoiling the xenon with other long-lived radioactive impurities [9]. The concentration of ^{83m}Kr in xenon is doped to the order of $\sim 10^{-15}$, which allows to observe the distillation process on a sub-ppt scale, due to the fact, that ^{83m}Kr should follow the same chemical behavior concerning the distillation process as the other krypton isotopes. The decay products (conversion- and Auger electrons as well as γ-rays) are reacting with the xenon and produce scintillation light of 171 nm wavelength [10] that can be detected using special photomultiplier tubes with optimized quantum efficiency for that region. The rate on the photomultiplier tubes is proportional to the concentration of ^{83m}Kr in the xenon.

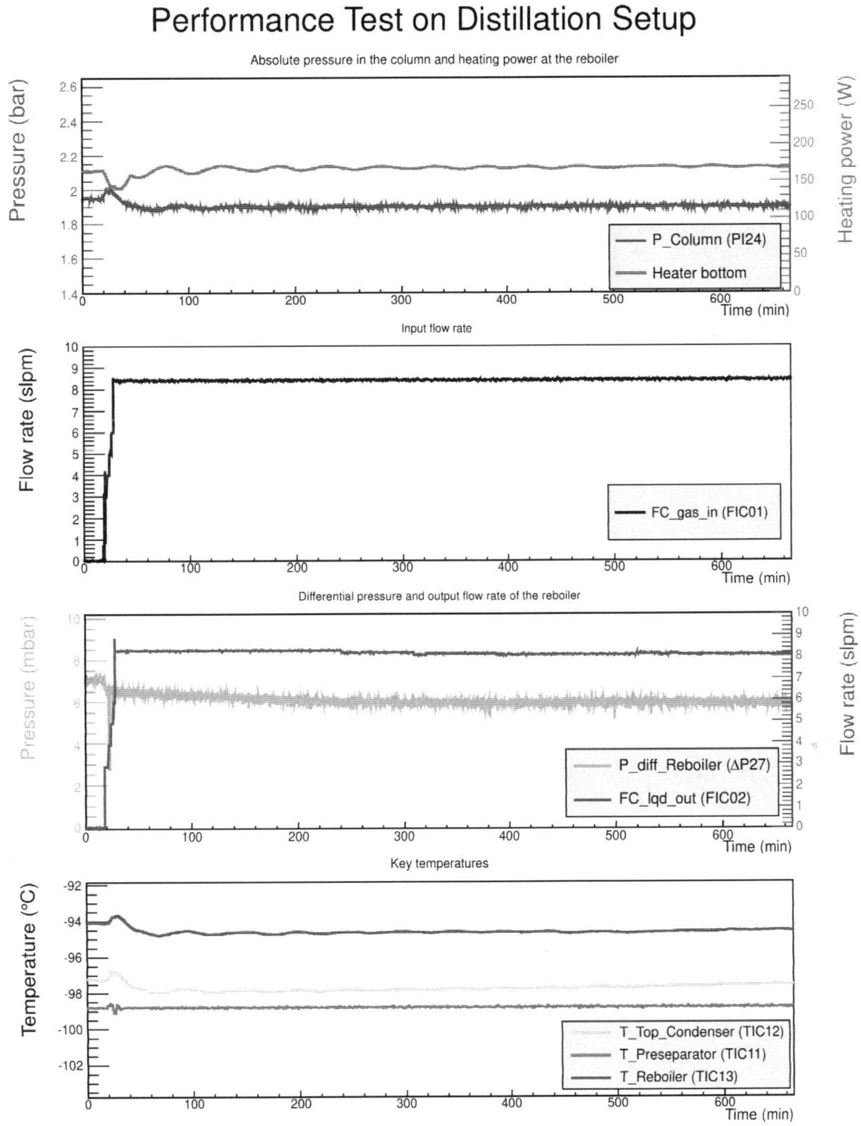

Figure 5. Performance test on the Phase-1 distillation plant. These plots show the key parameters of the distillation setup during one longterm distillation test of 11 hours under the design parameters.

The technical realization of these ^{83m}Kr-decay detectors is shown in figure 4. The performance of these detectors is presented in a dedicated paper [11].

5. Performance studies

In the following, the first performance studies will be presented. After the cleaning procedures, the column is cooled down by starting the coldhead with a set-point temperature, allowing the xenon to liquefy at the top. As consequence, the liquid droplets stream down, cooling the whole column from top to bottom. By adding more xenon, a liquid reservoir is build up in the reboiler, which is later needed to perform the distillation. The total load inside the column is ~5 kg. The regulation during the distillation process is designed to keep the pressure inside the column constant by varying the heater power using a PID regulation algorithm.

In figure 5 the key parameters for a distillation test of 11 hours are presented. The distillation process started after 20 minutes, when the input flow (FIC01) is slowly increased to the design value of 8.4 slpm (standard liter per minute). The liquid out flow at the bottom is also increased to keep the mass balance of incoming and outgoing fluxes of the system equalized. The system needs about 100 minutes to get to the equilibrium and running stable for several hours. The pressure in the system is stable at the set point for this measurement of 1.9 bar.

It has been demonstrated so far, that the Phase-1 column is thermodynamically stable under the designed flow- and pressure parameters. The investigation of the separation efficiency started recently using the rga system, and the ^{83m}Kr tracer method.

Acknowledgments

Different aspects of theses experiments have been funded by DFG-Großgeräte, BMBF and Helmholtz Alliance for Astroparticle physics (HAP). We would also like to acknowledge Ondrej Lebeda from the Nuclear Physics Institute, Academy of Science of the Czech Republic, Prague for kindly providing the ^{83}Rb generator for the ^{83m}Kr tracer technique.

References

[1] Ade P A R *et al* (Planck Collaboration) 2013, Planck 2013 results. XVI. Cosmological parameters, arXiv:1303.5076v3 [astro-ph.CO].

[2] Jungmann G *et al* 1996, Supersymmetric dark matter, *Phys. Rept.* **267** 195-373.

[3] Aprile E *et al* (XENON100) 2012, The XENON100 dark matter experiment, *Astroparticle Physics* **Volume 35**, Issue 9, 573-590.

[4] Aprile E *et al* (XENON100) 2012, Dark matter results from 225 live days of XENON100 data, *Phys. Rev. Lett.* **109**, 181301.

[5] Akerib D S *et al* (LUX Collaboration) 2014, First results from the LUX dark matter experiment at the Sanford Underground Research Facility, *Phys. Rev. Lett.* **112** 091303, arXiv:1310.8214 [astro-ph.CO].

[6] McCabe W L, Smith J C and Harriott P, 2005, Unit Operations of chemical engineering, 7th edition, McGraw-Hill international edition, p. 652

[7] Dobi A *et al* 2011, Detection of krypton in xenon for dark matter applications, *Nucl. Instrum. Methods* **A 665** 1-6.

[8] Brown E, Huhmann C, Kettling H, Rosendahl S and Weinheimer C, 2013, In situ measurements of krypton in xenon gas with a quadrupole mass spectrometer following a cold-trap at a temporarily reduced pumping speed, *JINST* **8** P02011.

[9] Hannen V *et al* , 2011, Limits on the release of Rb isotopes from a zeolite based 83mKr calibration source for the XENON project, *JINST* **6** P10013.

[10] Takahashi T *et al* , 1983, Emission spectra from Ar-Xe, Ar-Kr, Ar-N_2, Ar-CH_4, Ar-CO_2 and Xe-N_2 gas scintillation proportional counters, *Nucl. Instrum. Methods* **205** 591-596.

[11] Rosendahl S *et al* , 2014, A novel ^{83m}Kr tracer method for characterizing xenon gas and cryogenic distillation systems, *JINST* **9** P10010.

Journal of Physics: Conference Series **564** (2014) 012007 doi:10.1088/1742-6596/564/1/012007

Impulse Plasma In Surface Engineering - a review

K Zdunek[1,2 *], K Nowakowska-Langier[2], R Chodun[1], S Okrasa[1], M Rabinski[2], J Dora[3], P Domanowski[4] and J Halarowicz[5]
[1]Warsaw University of Technology, Woloska 141, 02-507 Warsaw, Poland,
[2]National Centre for Nuclear Research, Andrzeja Soltana 7, 05-400 Otwock, Poland
[3]DORAPS, Wilczycka 8, 51-361 Wilczyce, Poland
[4] University of Technology and Life Sciences in Bydgoszcz, Al. Prof. Kaliskiego 7, 85-796 Bydgoszcz, Poland
[5]BOHAMET, Torunska 2, 86-005 Biale Blota, Poland

E-mail: zdunek@inmat.pw.edu.pl

Abstract. The article describes the view of the plasma surface engineering, assuming the role of non-thermal energy effects in the synthesis of materials and coatings deposition. In the following study it was underlined that the vapor excitation through the application of an electric field during coatings deposition gives new possibilities for coatings formation. As an example the IPD method was chosen. During the IPD (Impulse Plasma Deposition) the impulse plasma is generated in the coaxial accelerator by strong periodic electrical pulses. The impulse plasma is distributed in the form of energetic plasma pockets. Due to the almost completely ionization of gas, the nucleation of new phases takes place on ions directly in the plasma itself. As a result the coatings of metastable materials with nano-amorphous structure and excellent adhesion to the non-heated intentionally substrates could be deposited. Recently the novel way of impulse plasma generation during the coatings deposition was proposed and developed by our group. An efficient tool for plasma process control, the plasma forming gas injection to the interelectrode space was used. Periodic changing the gas pressure results in increasing both the degree of dispersion and the dynamics of the plasma pulses. The advantage of the new technique in deposition of coatings with exceptionally good properties has been demonstrated in the industrial scale not only in the case of the IPD method but also in the case of very well known magnetron sputtering method.

1. The role of electric energy in nonconventional synthesis of materials

The use of plasma in vapour deposition methods of surface engineering is a consequence of the efforts to increase the efficiency of the process. This approach concerns a typical technological aspect of the process (for example, obtaining vapours from a solid phase through the use of an arc discharge instead of thermal evaporation or the possibility of sputtering dielectric targets by applying rf power).

Another tendency seems to be even more important: Application of plasma is providing new possibilities of influencing material parameters, such as morphology (e.g., achieving a greater dispersion of the crystallites, or "designing" the mixing zone at the interfacial boundary) or to form metastable phases such as DLC, c-BN etc. even in a relatively low temperature of the substrate.

---

* Corresponding author: Krzysztof Zdunek

The role of the particle energy during the formation of film morphology together with the substrate temperature is reflected in the well known film morphology models of Movchan-Demchishin (M-D) [1], Thornton [2], Messier [3] or Anders [4]. An underlining assumption of these models was to generalize the relationship between energy conditions of layer synthesis and the morphology of these layers, especially in view of the possibility to obtain fine crystalline structures instead of columnar structure.

In this context attention should be paid to the replacement of the pressure effect in the Thornton model by kinetic energy of particles in Messier and Anders models. The M-D model refers to the thermal evaporation and does not consider the role of pressure in the creating the morphology of the layers, while the Thornton model refers to magnetron sputtering when the gas under the pressure of about 10^0-10^1 Pa plays a crucial role. However, the pressure in the second model represents a measure of the nonthermal kinetic energy of the vapour phase particles. In this way, the linear M-D model with the layer structure dependent solely on the temperature is transformed into the Thornton model in which the film structure is described in a two-dimensional space (T, p), reflecting thereby the dependence of the structure on the gas phase particles (plasma) kinetic energy.

The role of the plasma in the development of the phase composition of a layer material in surface engineering with gaseous methods is vividly reflected by spectacular example of the DLC layers synthesis. It seems that the importance of the electric energy in the synthesis of carbon-linked sp^3 hybridization is not in doubt. This problem was launched with the publication of Aisenberg and Chabot [5]. It is interesting that in the cited work the authors restricted themselves to the kinetic energy obtained by the ionized atoms during their acceleration by the electric field, and the interaction of these ions with the substrate where nucleation of a "diamond" carbon phase occurred.

A different and original view of the DLC synthesis has been expressed by Sokolowski and Sokolowska [6,7]. These authors substantiated the formation of sp^3 bonding by carbon phase nucleation on plasma ions. The essential potential energy is supplied to the nuclei of this stage as a result of inelastic collisions with electrons. The role of the potential energy seized from the electric field applied in the nucleation zone has been demonstrated for example in [8] and [9] for the synthesis of an aluminium oxide and boron nitride, respectively. In this second paper taking the electric field into account it was explicitly proposed to define something like a new state parameter related to the potential energy of the plasma particles which, along with pressure and temperature determines the phase composition of the BN layers.

2. IPD – Impulse Plasma Deposition method

In the 1980's in the Faculty of Materials Science of Warsaw University of Technology the Impulse Plasma Deposition (IPD) method has been proposed. This method has no counterpart among other contemporary methods of surface engineering. As already mentioned above, it is based on the assumption that homogeneous nucleation takes place on plasma ions which have been formed by inelastic collisions with electrons.

The use of pulse plasma generated in a coaxial accelerator during short discharge in gas under reduced pressure is most favourable to activate this mechanism. Such conditions were created due to discharge of the capacitor with a capacity of the order of 10^{-4} F, cyclically loaded to a voltage of 10^3 V with a preset frequency of 10^{-1}-10^0 Hz, when peak current value reaches the order of 10^5 A.

The lifetime of the resulting plasma is of the order of 10^{-4} s. Plasma is emitted from the coaxial accelerator in the form of a plasmoid, which is accelerated by the magnetic pressure to speeds of about 10^5 ms^{-1}. During the interaction of the plasmoid with the unheated surface of the substrate a short (of the order 10^{-3} s) thermal pulse with a peak temperature of about 2000 K [10] acts on the substrate surface. This heat is dissipated in the substrate material with a velocity in the order of 10^6 Ks^{-1} which is determined by the thermal conductivity of the substrate. In addition to the thermal effect the plasmoid interaction with the substrate surface causes subplantation of plasma constituents, which favours the formation of the mixing zone at the interface.

The essential application-related features of the pulse plasma generated in the manner described above were connected with nearly complete ionization and thermodynamic nonequilibrium [11,12]. In such plasma, the probability of inelastic electron collision with the heavier plasma components, including the formation of critical nuclei in the plasma, is particularly high and therefore efficiency of energy exchange between the plasma and the condensed phase formed in plasma is also high.

It should be noted that in the impulse plasma generated during capacitor discharge (as a source of electrical energy) it is possible to induce discharges at voltages which are significantly lower than those described by the Paschen curve [13]. In this way a particularly large amount of energy can be supplied in a relatively easy way and, hence, the internal energy of the gas environment increased to an extent which is impossible to reach by other means. As a result, the mechanism of IPD layer synthesis comprises the following specific elementary phenomena [14]:

1. Generation of practically completely ionized nonequilibrium plasma by high-energy pulse discharge in gas under reduced pressure,
2. Acceleration of the plasmoid with magnetic pressure thus increasing the kinetic energy of the plasma,
3. Nucleation in plasma on ions,
4. Dynamic interaction of the plasmoid with the surface of the unheated substrate, leading to a strong thermal activation of the surface and to subplantation,
5. A cluster growth mechanism, consisting of a limited coalescence of nanoaggregates.

IPD method, although it is still somewhat a niche technique in plasma surface engineering, has demonstrated however its applicability in the synthesis of layers composed of such materials like DLC [15, 16], c-BN [9], Al_2O_3 [8] as well as of multi-component metallic alloys of MCr Al(Y)-type [16]. Industrial use of IPD method for depositing TiN layers on cutting tools in one of the largest Polish machinery works [18] was a spectacular illustration of technique potential. One should keep in mind that, according to our knowledge, this is the only method of plasma surface engineering which allows to obtain nanocrystalline wear resistant layers on unheated and unbiased substrates.

Recently a similar kind of pulse plasma generation was also used in magnetron sputtering technology [19 - 21], namely in form of HIPMS (High-Power Impulse Magnetron Sputtering). It is worth to underline that the essential features of the HIPIMS process, i.e. inducing the gas discharge under reduced pressure, using the capacitor as the source of energy, as well as taking the advantages of high degree of plasma ionization and its nonequilibrium state are similar to the assumptions of the IPD method. The practical achievements of the HIPMS technology application are connected first of all with a significant reduction of substrates temperature and a dramatic improvement of the morphology of the deposited layers, in particular the reduction of the tendency of structure columnarization.

3. Gas injection as a tool of control of plasma process during the coatings deposition

In 2011, Zdunek et al. [22] proposed a novel variant of plasma surface engineering processes, namely to control the plasma process through the use of pulse variable concentration of the working gas. The pressure of the gas injected with a pulse-valve having a low mechanical inertia cyclically oscillated between a value of the order of 10^{-4}-10^{-5} Pa and a value of the order of 10^{-1}-10^{-2} Pa.

Work in this regime allows preventing or triggering the electrical discharge in the gas. This idea has been practically proved in the IPD accelerator, previously equipped with continuous dosing of the working gas (standard IPD).

The following assumption was applied. Operating a valve with the opening time of the order of 10^{-3} s and a frequency of the order of 10^{-1} Hz (highly dependent on the pumping system efficiency relation to the volume of the vacuum chamber) leads to the generation of plasma in the coaxial accelerator during the time of pressure rise in the interelectrode space. Since the plasma generation takes place during the pressure increase, the rise of injected gas mass is followed by the current value increase due to the growing concentration of the gas undergoing ionisation. This synchronization should result in strong differentiation of the plasmoid velocities at different stages of its development, thus promoting a greater degree of plasma nonequilibrium in the IPD method. As before valve opening the mean free path of the gas particles many times exceeds the linear dimensions of the vacuum chamber (for a pressure of 10^{-4} Pa the mean free path is of the order of 10^{1} m) the plasmoid accelerated with magnetic pressure does not collide with the cold gas in the vacuum chamber volume after leaving the interelectrode space. This and the lower average gas density of the plasmoid in comparison to that generated under continuous gas dosing should favour reducing the dissipation of the kinetic energy of plasma particles during collisions.

This probably reduces the tendency to aggregate nuclei by plasma ion nucleation in comparison to the standard IPD process. As a result, in the ion flux hitting the substrate surface the portion of atomic ions increases and that of cluster ions decreases. Consequently, the whole layer growth mechanism is changed in such a manner that atomic mechanism play are larger role and nuclei formation also on the substrate surface takes place.

The research hypothesis presented above has been verified during subsequent studies. Already published results show that:

- A large increase of TiN coated SW7M HSS tool life, reaching up to 1600% [23], was observed
- Morphologically, the TiN layer represents a nanograin structure, in contrast to nanoglobular structure in the standard IPD [23].

Unpublished research so far carried out proves that:

- The electron temperature of the plasmoid spreading in a vacuum chamber reaches 10 eV at the substrate (Langmuir probe, computer simulations).
- The plasmoid is almost completely ionized (OES).
- Multiple ionized titanium and nitrogen are present within plasmoid (OES).
- At the surface of the iron substrate influenced by the plasmoid exited atoms of iron are observed (OES). We conclude from this that intensive sputtering and/or evaporation of iron from this surface takes place.
- In the interfacial zone a considerable enrichment of the substrate with the titanium is observed (HRTEM, SIMS),
- Locally the increase of the TiN layer has explicit hetero epitaxial character (HRTEM)
- Adhesion of the TiN layers to the SWAM steel substrate is very good, corresponding to a critical load of 80-100 N (scratch test method).

The use of the mode described above during magnetron sputtering process could be an interesting example of pulse gas concept application in the plasma surface engineering. In [22] preliminary results are presented for the synthesis of AlN layers using this solution (GIMS - Gas Injection Magnetron Sputtering). As a result of pulsed dosing of the working gas, the AlN layers on the surface of monocrystalline silicon substrates do not exhibit a characteristic columnar structure. The growth rate of the layers, however, was decreased by a factor of three compared to the process with continuous dosing of gas.

Technical usability of GIMS was extremely positively verified in industrial conditions. The GIMS concept was used in a multimagnetron facility for the deposition of Ti and TiO_2 layers on large-size unheated windows of 2000x3000 mm, utilizing linear magnetrons with the length of 2200 mm. The Ti layers obtained were distinguished by a much better adhesion (qualitatively tested using adhesive tape) than layers produced with continuous dosing of the working gas. A spectacular result was obtained in the case of TiO_2 layers produced on ceramic unheated tiles. Abrasion studies of the tile surface coated with TiO_2 proved class II abrasion resistance, enabling the use of coated tiles for lightly loaded floors (e.g. in operating rooms).

4. Conclusions

Non-thermal energy can be utilised in gaseous surface engineering methods as a specific substitute of thermal energy. The synergy of both types of energy in the synthesis of layers is evident and creates new possibilities for control of the process synthesis. Non-thermal energy of the particles is created by the application of electric power in plasma surface engineering. Two cases have to be discriminated, one is associated with the kinetic energy of the particles, the other with their potential energy as represented by ionisation of excitation of particles. The latter aspect seems to be significant because of providing impact on the solid phase composition of the synthesis products occurring in the plasma itself. As a result, it is possible to nucleate condensed phases in the plasma which is otherwise characterized by high barriers of nucleation.

The recently introduced innovative mode of process control in the plasma surface engineering methods through the use of dynamically changing pressure of working gas opens new perspectives in the development of these methods. It seems that the use of this mode allows to increase the efficiency of electric energy transfer to the gas source, which leads for example to an increase in the ionization degree or the electron temperature in comparison to a plasma generated under the same conditions but with continuous dosing of the working gas. Probably the kinetic energy dissipation by collisions of plasma particles decreases, which positively affected the status of layer/substrate interface. This yields better application properties of the layers.

Acknowledgement

This work was financially supported by the National Centre for Research and Development under the contract NR15-0002-10/10

References

[1] Movchan B and Demchishin A 1969 *Fiz. Miet. Mietall.* **28** 653
[2] Thornton J A 1974 J *Vac. Sci. Technol.* **11** 666
[3] Messier R, Giri A P and Roy R A 1984 *J. Vac. Sci. Technol* **A2** **2** 500
[4] Anders A 2010 *Thin Solid Films* **518** 4087
[5] Aisenberg S and Chabot R 1971 *J. Appl. Phys.* **42** 2953
[6] Sokolowski M 1981 *J. Cryst. Growth* **54** 519
[7] Sokolowski M and Sokolowska A, 1982 *J. Cryst. Growth* **57** 185
[8] Sokolowski M, Sokolowska A, Michalski A and Ziolkowski Z 1981 *J. Cryst. Growth* **52** 274
[9] Sokolowska A and Wronikowski M 1986 *J. Cryst. Growth* **76** 511
[10] Romanowski Z and Wronikowski M 1992 *J.Mater.Scien.* **27** 2619
[11] Kulakowska B, Zyrnicki W and Sokolowska A 1990 *J.Less-Comm.Met.* **A** **166** 81
[12] Kulakowska-Pawlak B 2009 *Plasma Sources Sci.Techn.* **B** **18** 1
[13] Paschen F 1889 *Annalen der Physik und Chemie* **273** 69

[14] Zdunek K 2007 *Surf. Coat. Technol.* **201** 4813
[15] Szmidt J, Sokolowska A, Zdunek K and Mitura S 1991 *Diamond Relat. Mater.* **1** 588
[16] Sokolowska A, Olszyna A, Michalski A and Zdunek K 1991 *Surf. Coat.Technol.* **47** 144
[17] Zdunek K and Grigoriew H 1995 *J. Mater. Sci.* **30** 4479
[18] Zdunek K 1995 *Surf.Coat.Techn.* **74-75** 949
[19] Kouznetsov V, Macak K, Schneider J M, Helmersson U and Petrov I 1999 *Surf.Coat.Technol.* **122** 290
[20] Macak K, Kouznetsov V, Schneider J, Helmersson U and Petrov I 2000 *J. Vac. Sci. Technol.* **A18** 1533
[21] Helmersson U, Lattemann M, Bohlmark J, Ehiasarian A P and Gudmundsson J T 2006 *Thin Solid Films* **513** 1
[22] Zdunek K, Nowakowska-Langier K, Dora J and Chodun R 2013 *Surf.Coat.Techn.* **228** 367 This work was first presented orally at AEPSE 2011 in Dalian, China, in 2011
[23] Zdunek K, Nowakowska-Langier K, Chodun R, Kupczyk M and Siwak P *Vacuum* **85** 514